AF774837

BIOCHEMISTRY RESEARCH TRENDS

APIGENIN AND NARINGENIN

NATURAL SOURCES, PHARMACOLOGY AND ROLE IN CANCER PREVENTION

BIOCHEMISTRY RESEARCH TRENDS

Additional books in this series can be found on Nova's website
under the Series tab.

Additional e-books in this series can be found on Nova's website
under the e-book tab.

BIOCHEMISTRY RESEARCH TRENDS

APIGENIN AND NARINGENIN

NATURAL SOURCES, PHARMACOLOGY AND ROLE IN CANCER PREVENTION

NILUS M. STACKS
EDITOR

nova publishers
New York

Copyright © 2015 by Nova Science Publishers, Inc.

All rights reserved. No part of this book may be reproduced, stored in a retrieval system or transmitted in any form or by any means: electronic, electrostatic, magnetic, tape, mechanical photocopying, recording or otherwise without the written permission of the Publisher.

For permission to use material from this book please contact us:
nova.main@novapublishers.com

NOTICE TO THE READER

The Publisher has taken reasonable care in the preparation of this book, but makes no expressed or implied warranty of any kind and assumes no responsibility for any errors or omissions. No liability is assumed for incidental or consequential damages in connection with or arising out of information contained in this book. The Publisher shall not be liable for any special, consequential, or exemplary damages resulting, in whole or in part, from the readers' use of, or reliance upon, this material. Any parts of this book based on government reports are so indicated and copyright is claimed for those parts to the extent applicable to compilations of such works.

Independent verification should be sought for any data, advice or recommendations contained in this book. In addition, no responsibility is assumed by the publisher for any injury and/or damage to persons or property arising from any methods, products, instructions, ideas or otherwise contained in this publication.

This publication is designed to provide accurate and authoritative information with regard to the subject matter covered herein. It is sold with the clear understanding that the Publisher is not engaged in rendering legal or any other professional services. If legal or any other expert assistance is required, the services of a competent person should be sought. FROM A DECLARATION OF PARTICIPANTS JOINTLY ADOPTED BY A COMMITTEE OF THE AMERICAN BAR ASSOCIATION AND A COMMITTEE OF PUBLISHERS.

Additional color graphics may be available in the e-book version of this book.

Library of Congress Cataloging-in-Publication Data

ISBN: 978-1-63463-987-3
Library of Congress Control Number: 2014960320

Published by Nova Science Publishers, Inc. † New York

CONTENTS

PREFACE

Apigenin is one of the most bioactive naturally occurring flavonoids, found in fruits, plant-derived beverages, vegetables, wheat sprouts, and some seasonings. It has been found to possess anti-oxidant, anti-inflammatory, and anti-cancer properties. Naringenin is one of the most studied flavonoids abundant in citrus plants. Naringenin and their derivatives are known to exhibit strong anti-oxidant potential along with plenty of protective effects for the improvement of human health. This book discusses the natural sources, pharmacology and role in cancer prevention of the flavonoids, apigenin and naringenin.

Chapter 1 - Apigenin is a naturally occurring plant flavone with prominent antioxidant and anti-inflammatory properties. Although apigenin has the potential to be a promising molecule also for cancer treatment, its delivery to the body requires suitable dosage form design due to physicochemical characteristics that lead to poor bioavailability. It is classified as a BCS (Biopharmaceutical Classification System) II drug with low water solubility and high lipophilicity.

Improving the solubility of apigenin is crucial, various formulation approaches have recently been employed e.g., nanocrystals have increased dissolution rate and antioxidant activity, while polymeric micelles increased the solubility to 148 times higher than that crude apigenin and showed enhanced cytotoxicity on hepatoma cells. PLGA encapsulated apigenin nanoparticles cause mitochondrial apoptosis and even DNA targeting could be achieved in skin cancer.

The chapter reviews current approaches to the formulation of drug delivery systems containing apigenin.

Chapter 2 - Apigenin (4′, 5, 7-trihydroxyflavone) is a flavone, naturally occurring in many fruits and vegetables, such as oranges and grapefruits.

Apigenin has been progressively studied in the last decades for its potential use as anticancer agent, since its attractive anti oxidant, anti inflammatory properties associated to its low toxicity. In addition apigenin has been demonstrated to be non mutagenic.

However, although apigenin has proved to counteract tumour growth both in *in vivo* and *in vitro* studies, the mechanisms underlying remains largely unknown.

Apigenin exhibits a broad anti cancer effect, acting on protein kinases it modulates cell cycle and apoptosis, thus inhibiting cancer cell proliferation, moreover Apigenin might modulate carcinogens metabolism and disposition. Furthermore Apigenin counteracts tumour spreading, thorugh inhibition of cell motility and metastatization.

Metastasis is the final step of tumor progression and it represents one of the most challenging characteristics of malignant transformation and the main cause of cancer-related deaths. It is referred to as the diffusion of cancer cells outside the primary site of origin, through a complex process that encompasses cell adhesion, migration and immune escape.

Several studies have demonstrated apigenin ability in reducing cell motility, migration and invasion onto ontogenic different types of cancer. The molecular mechanisms underlying this effect are not completely understood, even if the modulation of MAPK and PI3K pathways are surely involved.

In this chapter the results of both *in vitro* and *in vivo* studies that evaluated the anti metastatic potential of apigenin will be summarized aiming at elucidate the molecular aspects involved. Moreover a comprehensive analysis of the methods and *in vivo* and *in vitro* models used in this field will be done.

Chapter 3 - Few last decades witnessed increasing research reports on flavonoids because of their wide range of applications. Flavonoids have been used in completely different fields ranging from the food industry to medicine. Apigenin is one of the several flavonoids being researched and reported to have relatively low toxicity compared to other flavonoids. The main limitation in exploration of apigenin's effects is mainly due to the low water solubility which causes many formulation issues and hinders bioavailability of compound. Successful attempts are made to incorporate apigenin in various nanoformulations like micelles, polymeric nanoparticles and lipid nanocapsules. Several fold solubility enhancement of apigenin is mainly observed when nanocrystal technology is used. This chapter takes extensive overview of nanotechnology based apigenin encapsulated drug delivery systems and their performance *in vitro* and *in vivo*.

Chapter 4 - Inflammation is a key player in numerous human diseases. Several evidences support the relationship between inflammation and cancer. This is based on the finding that tumors often arise at sites of chronic inflammation. Natural flavonoids have been used to treat various inflammatory diseases. In particular, apigenin is a flavonoid that is widely distributed in dietary plants, especially in parsley, thyme, peppermint, olives, and herbs like chamomile. This review summarizes the current knowledge on the effects of apigenin on inflammation process, with a focus on the cellular and molecular mechanism involved. In this sense, different molecular and cellular targets are analyzed. Few clinical trials have described the relevance of apigenin in human diseases; the need for further controlled intervention trial will be discussed. Finally, taking into account its poor aqueous solubility leading to a poor absorption in gastrointestinal tract, novel strategies to improve its solubility and bioavailability, such as polymeric micelles, ethosomes, liposomes, will be shown.

Chapter 5 - In this review, inhibitory activity against mammalian DNA polymerases (pols) by ten major plant flavonoids was investigated. Apigenin (4′,5,7-trihydroxyflavone) was the most potent inhibitor of pols among the compounds tested. Apigenin did not affect the activity of plant (cauliflower) pol α or prokaryotic pols, such as the Klenow fragment of pol I from *Escherichia coli*, *Taq* pol and T4 pol. Apigenin did not inhibit the activities of other DNA metabolic enzymes, such as mouse IMP dehydrogenase (type II), T4 polynucleotide kinase and bovine deoxyribonuclease I. Apigenin also did not influence direct binding to double-stranded DNA as determined by thermal transition analyses. The ten major flavonoids were investigated as to whether they could prevent proliferation of human colon HCT116 carcinoma cells. Suppression of growth of human cancer cells showed the same tendency as inhibition of activity of mammalian pols. Apigenin showed the strongest suppression of proliferation of human cancer cells. These results suggest that the decrease of proliferation of cancer cells by flavonoids containing apigenin could be based on the inhibition of cellular pols. Apigenin did not suppress the growth of human normal cells (human dermal fibroblasts), and induced apoptosis in human cancer (HCT116) cells. Therefore, apigenin could be used as an anti-cancer agent, and apigenin-enriched fractions could be used in functional foods for cancer prevention.

Chapter 6 - Apigenin (4',5,7-trihydroxyflavone), a naturally occurring flavone, is widely distributed in many fruits and vegetables, such as parsley, chamomile, celery and kumquats. In recent years, apigenin has been increasingly recognized as a cancer chemopreventive agent. The chemopreventive aspects of apigenin have been evaluated both *in vitro* and *in vivo*. Several studies in different *in vitro* and *in vivo* systems have shown that apigenin possesses the capacity to regulate response to oxidative stress and DNA damage, suppress inflammation and angiogenesis, inhibit cell proliferation, and induce autophagy and apoptosis. One of the most obvious mechanisms of apigenin is to induce p53 tumor suppressor proteins at the translational levels, followed by $p21^{WAF1/CIP1}$ induction. Another role of apigenin in chemoprevention is induction of autophagy in several human cancer cells, including HCT116 human colon cancer cells. The authors recently reported that apigenin suppressed growth of HCT116 cells in a concentration-dependent manner by causing G2/M cell cycle arrest. In addition, apoptosis and autophagy were induced in apigenin-treated cells. Inhibition of autophagy by 3-methyladenine enhanced the apoptosis induced by apigenin through activation of pro-caspases-8, -9, and -3 and cleavage of poly(ADP-ribose) polymerase.

These results provide evidence that inhibition of autophagy may be an effective way to improve the chemotherapy of anticancer agents against human colon cancer. Therefore, this chapter discusses the induction of apoptosis and autophagy as a mechanism of chemoprevention by apigenin.

Chapter 7 - Apigenin (4',5,7-trihydoxyflavone, $C_{15}H_{10}O_5$) is one of the most bioactive naturally occurring flavonoids, found in fruits, plant-derived beverages, vegetables, wheat sprouts, and some seasonings. It has been found to possess anti-oxidant, anti-inflammatory, and anti-cancer properties. Research on the effects of apigenin at the cellular and molecular levels have shown positive results against breast, cervical, colon, lung, hematological, prostate, ovarian, thyroid, gastric, pancreatic, endometrial, liver, and skin cancers. Recent findings have also indicated that apigenin may be an alternative medication for prevention and/or therapy of allergic asthma and Alzheimer's disease. There is little evidence to date to suggest that apigenin promotes adverse metabolic effects when consumed as part of a normal diet. This chapter describes the natural sources, chemical structure, biological properties, and role of apigenin in cancer prevention. Several lines of evidence suggest that apigenin holds much potential to be developed as a cancer chemopreventive agent despite its low bioavailability.

Chapter 8 - A primary function of gastrointestinal epithelial cells is to provide a physical barrier to the diffusion of pathogens, toxins, and allergens from the external environment into the tissues. Defects in the barrier function resulting in the permeation of luminal inflammatory substances induce an abnormally robust inflammatory response. It has been reported that barrier defects play a crucial role in the pathogenesis of various intestinal diseases, such as inflammatory bowel disease, irritable bowel syndrome, and food allergies. Therefore, the maintenance of the intestinal barrier is imperative for intestinal mucosal homeostasis and our health. The intestinal barrier is maintained through the interactions of several barrier components, including the adhesive mucous gel layer, the mucosal immune system, and intercellular tight junctions (TJs). The TJ is a multiple protein complex, which locates around the apical end of the lateral membrane of polarized epithelial cells and selectively regulates the paracellular passage of ions, molecules and water. Although the TJ structure has been elucidated on a molecular level in recent studies, less information is available on the roles of food factors on intestinal barrier regulation. Naringenin is a major polyphenol isolated from citrus fruits. Recently, the authors have shown that naringenin enhances and protects TJ barrier integrity in intestinal cells and a murine model of colitis. Accordingly, naringenin appears to participate in the physiological and pathological regulation of the intestinal TJs. This chapter presents the recent advances in the author's understanding of the naringenin-mediated regulation of the intestinal TJ barrier.

Chapter 9 - Naringenin is one of the most studied flavonoids abundant in citrus plants. Most of the flavonoids derived from plants are found to be highly beneficial to human health because of their intrinsic antioxidant potential. Recent research on these small molecules have expanded on their effects and the molecular mechanisms for their modes of action in the prevention of a wide range of diseases. Particularly, naringenin and their derivatives are known to exhibit strong anti-oxidant potential along with plenty of protective effects for the improvement of human health. Several *in vitro* and *in vivo* experimental results support their beneficial effects, however, the lack of adequate clinical studies have limited their use as therapeutics. In this chapter, the occurance and dietary sources of naringenin will be briefly discussed along with naringenin biosynthesis pathway in plants. Moreover, the chapter will focus on discussing several protective effects of the compounds and their derivatives against cancer, obesity, diabetes, hypertension, hyperlipidemia, neuro-degenerative disorders, and other metabolic syndromes.

In: Apigenin and Naringenin
Editor: Nilus M. Stacks
ISBN: 978-1-63463-987-3
© 2015 Nova Science Publishers, Inc.

Chapter 1

DRUG DELIVERY APPROACHES FOR APIGENIN: *A REVIEW*

Zsófia Edit Pápay[1], Emese Balogh[1], Mohammed Gulrez Zariwala[3], Satyanarayana Somavarapu[2] and István Antal[1,*]
[1]Department of Pharmaceutics, Semmelweis University, Budapest, Hungary
[2]Department of Pharmaceutics, UCL School of Pharmacy, London, UK
[3]University of Westminster, London, UK

ABSTRACT

Apigenin is a naturally occurring plant flavone with prominent antioxidant and anti-inflammatory properties. Although apigenin has the potential to be a promising molecule also for cancer treatment, its delivery to the body requires suitable dosage form design due to physicochemical characteristics that lead to poor bioavailability. It is classified as a BCS (Biopharmaceutical Classification System) II drug with low water solubility and high lipophilicity.

Improving the solubility of apigenin is crucial, various formulation approaches have recently been employed e.g., nanocrystals have increased dissolution rate and antioxidant activity, while polymeric micelles increased the solubility to 148 times higher than that crude apigenin and showed enhanced cytotoxicity on hepatoma cells. PLGA encapsulated apigenin nanoparticles cause mitochondrial apoptosis and even DNA targeting could be achieved in skin cancer.

The chapter reviews current approaches to the formulation of drug delivery systems containing apigenin.

---

* E-mail: antal.istvan@pharma.semmelweis-univ.hu.

NATURAL SOURCES, INTAKE AND PHARMACOLOGY

Apigenin is a 4',5,7-trihydroxyflavone (Figure 1) belonging to a vast group of polyphenolic compounds called flavonoids. Flavonoids occur ubiquitously in the plant kingdom as secondary metabolites. They have a role in plant physiology, especially in pigmentation and flavor and also provide resistance against pathogens and insects. Six subclasses of flavonoids are distinguished - namely flavones, flavonols, flavanones, catechins, anthocyanidins and isoflavones - that have hydroxyl and phenolic groups variously attached to the common diphenylpropane structure (C6-C3-C6) (Ross and Kasum 2002). Generally one or more sugar components are linked to the aglycon molecule with O- or C-glycosidic bond in plants. Thus flavonoids are presented widely as glucosides in food but they are known to demonstrate incomplete absorption. They are abundantly distributed in fruits, vegetables and beverages with plant origin such as wine and tea. The average daily intake of flavonoids in a normal diet is approximately 1-2 g (Havsteen 2002). In the Hungarian population the total flavonoid intake was estimated to be 18.80 mg with only 0.58 mg apigenin (Lugasi, Hóvári et al. 2003). Many mechanisms of action have been identified for flavonoids, including antioxidant activity (Prochazkova, Bousova et al. 2011), changes in cellular signaling, apoptosis induction, anti-proliferation and anti-inflammation (Birt, Hendrich et al. 2001), thus suggesting their potential in cancer prevention. The use of pure flavonoids as a treatment in many common diseases such as cardiovascular and gastrointestinal disorders is increasing. Researchers have a growing interest in natural active ingredients to prevent or treat cancer, since most drugs currently available in the market have several disadvantages, being very toxic, highly inefficient or highly expensive. Unfortunately, the apparently effective concentrations of polyphenols in vitro are often higher than the absorbed amount measured in vivo.

Apigenin can be found in a variety of fruits and vegetables (Peterson and Dwyer 1998).

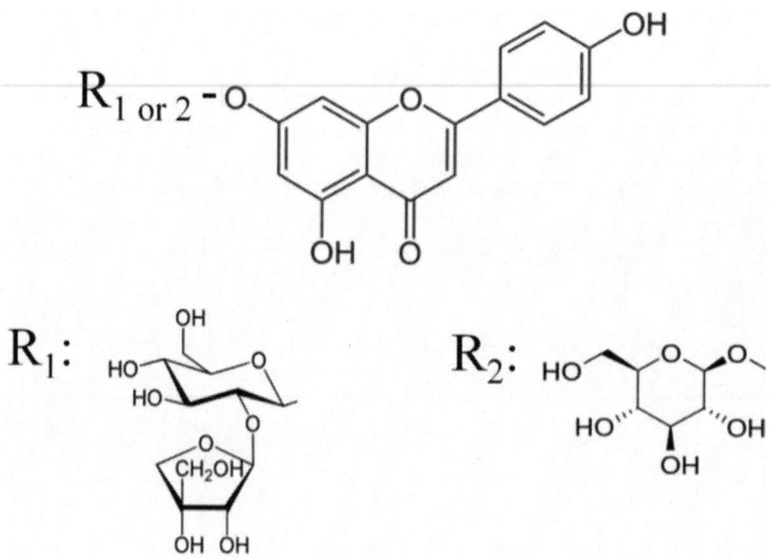

Figure 1. Chemical structure of apigenin and its glycosides.

The main sources of apigenin are celery, parsley, onion and chamomile, and it is present mainly in glycosides (Database 2014, May). However it occurs in a very low amount in the human diet (Meyer, Bolarinwa et al. 2006). It is recognized that apigenin has many pharmacological activities like free radical scavenging (Horvathova, Novotny et al. 2003, Škerget, Kotnik et al. 2005) and anti-inflammatory activities (Funakoshi-Tago, Nakamura et al. 2011, Choi, Islam). Furthermore, it has anticancer effects via modifying cell signaling pathways (Patel, Shukla et al. 2007, Shukla and Gupta 2010) in several tumor cell lines including skin (Wei, Tye et al. 1990, Tong, Van Dross et al. 2007, Abu-Yousif, Smith et al. 2008), breast (Yin, Giuliano et al. 2001), colon (Wang, Heideman et al. 2000, Chunhua, Donglan et al. 2013) and pancreas (Lefort and Blay 2013, Wu, Yu et al. 2014). Recent clinical studies suggest that a diet high in apigenin could potentially reduce the incidence of ovarian cancer in women (Gates, Vitonis et al. 2009).

All of these findings indicate that apigenin is a very promising drug candidate against cancer and several other diseases. However, the physicochemical and biopharmaceutical properties reveal limited clinical use highlighting a need to develop new drug delivery systems.

Physicochemical Properties and Bioavailability

Apigenin was classified as a drug belonging to BCS II. group (Biopharmaceutical Classification System) in a recent study (Zhang, Liu et al. 2012). This indicates high permeability but low solubility: the log P value is 2.87 (Li, Robinson et al. 1997), but the aqueous solubility is very poor either in water (1.35 μg/mL) (Li, Robinson et al. 1997) or in buffers. The highest solubility value was at pH=7.5 (2.16 μg/mL) (Zhang, Liu et al. 2012) leading to poor intestinal absorption. Several studies have reported that the solubility of flavonoids depends on temperature, nature of solvents, pH conditions, as well as the thermodynamic properties of the compounds which allow formation of hydrogen bonds with the surrounding solvent (Saidman, Yurquina et al. 2002, Tommasini, Raneri et al. 2004, Xiao, Shao et al. 2011).

The bioavailability of these molecules has been shown to be influenced by their chemical form in foods, the food matrix, and the consumer's microbial flora (Birt, Hendrich et al. 2001). Apigenin could be detected in human plasma following ingestion of apigenin-rich food, but in a very low amount. The first pass metabolism in the small intestine and the liver in conjunction with enterohepatic recycling may play an important role in the poor systemic bioavailability observed (Chen, Lin et al. 2003). Several other factors can cause low bioavailability, including low solubility and dissolution rate, poor membrane permeation, rapid metabolism and elimination. Very less is known about the fate of apigenin in the body.

Gastrointestinal tract is the main absorption site with both active and passive transport mechanism. This was confirmed with an intestinal transport study where apigenin could be detected in the whole intestine, with the main absorption site is probably being the duodenum. The mechanism might be passive transport in the ileum and colon segments with a concentration-independent permeability, while in both the duodenum and jejunum segments an active carrier-mediated transport is involved with concentration-dependent permeability behavior (Zhang, Liu et al. 2012). It seems likely that apigenin in natural form bound to β-glycosides (apiin, apigetrin) provides better bioavailability, as the absorption of other flavonoid glycosides is generally more efficient than the absorption of aglycones. (Hollman P C 1996). This result is more likely because glycosides are water soluble forms of the aglycon and more stable, (e.g., against heating such as cooking), thus more glycosides can get to the body (Nielsen, Young et al. 1999). However, the sugar group makes the molecule quite polar (log P < 0), and therefore hindering cellular uptake and passage through the cell membrane (Walle and Walle 2002, Walle, Hsieh et al. 2004). It was thought that flavonoids could not absorb through the small intestine because glycosides can only be hydrolyzed by a β-glucosidase enzyme of microbial flora in the colon (Pekić, Zeković et al. 1994). However, recently three types of human β-glucosidase enzymes were discovered; one of them capable of breaking the glycosidic bond and is located in the cells of the small intestine, liver and kidney (Asim 1993, Robert, Venkatakrishnan et al. 1993, Berrin, Czjzek et al. 2003). In blood circulation, apigenin binds to the main carrier protein – human serum albumin (HSA) – with hydrophobic and electrostatic forces. It is noteworthy that there are several type of binding mode between apigenin and HSA, but most of them are very instable (Yuan, lv et al. 2007).

Drug Delivery Approaches

In the last few decades the use of nutraceuticals to prevent chronic illnesses like diabetes, cardiovascular diseases and cancer has been rapidly emerging. These naturally occurring molecules possess antioxidant activity therefore can be effective against oxidative damage induced diseases. Various bioactive compounds from nutraceuticals and herbal medicines like flavonoids and other polyphenols have proven to be effective and have been applied with safe outcomes. Nevertheless, the exact dose and the complex mode of action are still unfolding areas (Nijveldt, van Nood et al. 2001). There is a need to investigate the influential factors on e.g., antioxidant activity during formulation development.

In a recent study the antioxidant activity influenced by extraction process was measured. It was concluded that temperature and grinding have negative effect on antioxidant activity and flavonoid content, moreover the glucosides have lower antioxidant activity than aglycone (Pápay and Antal 2014). However, their low bioavailability *in vivo* limits their usage in medical therapy (Nijveldt, van Nood et al. 2001). Researcher's attention has been drawn by nanotechnology which can change the pharmacokinetics and biodistribution thus improving the bioavailability and effectiveness (Huang, Yu et al. 2010). Controlled drug delivery systems makes targeted drug delivery possible thus leaving the healthy tissues unharmed and decrease the potential of side effects. This is extremely important in case of systematically administered chemotherapeutic agents that can cause severe side effects (Ferrari 2005, Merisko-Liversidge and Liversidge 2008). On the contrary, phytochemicals with anticancer properties have mild or negligible side effects. Several flavonoids including quercetin, luteolin and rutin were loaded into nanoparticles and have been found to be useful as nanomedicine (Nair, Sung et al. 2010). Solubility improvement of apigenin is crucial in order to achieve therapeutic effect as observed previously, and the delivery of apigenin requires appropriate pharmaceutical formulation to provide protection of the therapeutically active molecular form until delivery to the target tissue. In the last decades a number of drug delivery approaches for apigenin have been developed, including the use of nanotechnology.

Encapsulation into Inclusion Complexes and Colloidal Carriers

Encapsulation into inclusion complexes or colloidal carriers may serve as technologies for enhancement of solubility and bioavailability.

The molecular inclusion complexation can be achieved by using well known pharmaceutical solubilizers, namely cyclodextrins (CD). Natural CDs are consist of 6 (α), 7 (β), 8 (γ) glucopyranose units. These molecules are cyclic oligosaccharides made from starch by bacterial enzymatic conversion. They have truncated cone shape with hydrophobic cavity and hydrophilic rim due to the chair conformation of the units. They have the ability to form non-covalent dynamic inclusion complexes with poorly water soluble drugs thus increasing their aqueous solubility, stability and bioavailability, especially for BCS II. and IV. drugs (Loftsson and Brewster 1996), (Szente and Szejtli 1999). A number of data have been reported in the last years about the inclusion of plant materials (Pinho, Grootveld et al. 2014), but only one about the complexation of apigenin. Kim et al. investigated the aqueous solubility of apigenin with natural β-CD, its methylated (heptakis (2,6 di *O* methyl) β cyclodextrin, DM-β-CD) and hydroxylated (2-hydroxypropyl-β-cyclodextrin, HP-β-CD) derivates. The possible structure of complexation can be seen on Figure 2. The highest solubilization efficiency could be achieved with HP-β-CD: 11.5 fold increase with 2 mM HP-β-CD compare to water. They also conducted phase solubility studies, where 1:1 stoichiometric flavonoid/CD complexation can be assumed from the plotted phase solubility diagrams. Stability constant (K_C) values were the following: 827.6 with β-CD, 1038.6 with DM-β-CD and 4511.5 with HP-β-CD. The strongest complexation occur with HP-β-CD, possibly due to hydrogen bonds between apigenin and hydroxyl groups (Kim H 2008). The degree of hydroxylation affects the hydrogen bonding capacity, and thus interactions with membrane (Ollila, Halling et al. 2002).

For improving biological activity, apigenin can be encapsulated into liposomes or micelles, too (Figure 3).

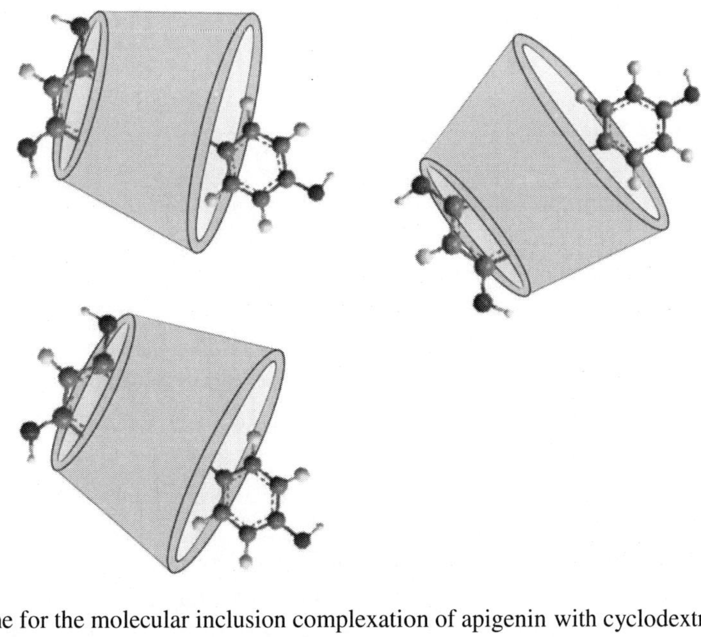

Figure 2. Scheme for the molecular inclusion complexation of apigenin with cyclodextrin.

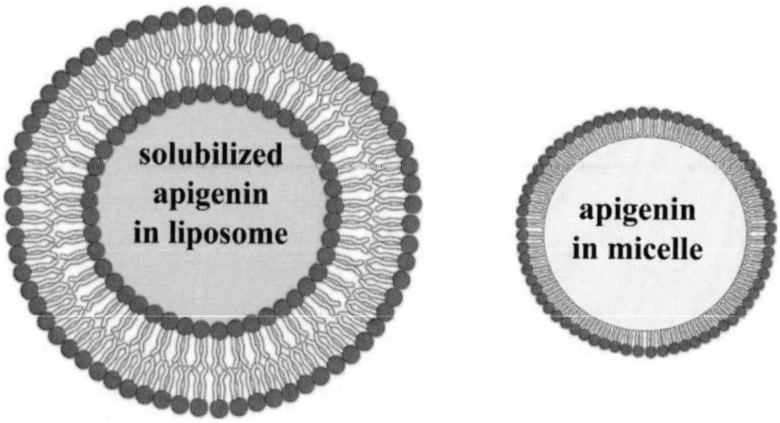

Figure 3. Encapsulation of apigenin into colloidal carriers.

Lipid nanocapsules are patented biomimetric carriers which have characteristic hybrid structure of polymer nanocapsules and liposomes.

Lipid nanocapsules (LNC) were applied to encapsulate apigenin with phase inversion method. Generally medium-chain triglycerides build up the oily core which is surrounded by mixture of lecithin and pegylated surfactant as a membrane. One of their advantages compared to liposomes the solvent free preparation process and longer stability, up to 18 months (Huynh, Passirani et al. 2009). The experiment was optimized by simplex lattice design experiment, for indexes drug loading and encapsulation efficiency was used. The optimal formulation resulted 1.26 ± 0.05% drug loading and 95.86 ± 0.38% encapsulation efficiency. The total apigenin concentration was 5.88 mg/mL with well dispersed, spherical in shape nanocapsules with average particle size 46.1 nm and -28.18 mV zeta potential. The *in vitro* release behavior showed two phase dynamics process and *in vivo* MTT assay showed enhanced antiproliferative activity. The study concluded that lipid nanocapsules are potential carrier for apigenin to improve its solubility and biological activity (Ding, Chen et al. 2013).

Topical Delivery

One of the studies was to investigate the influence of vehicle, distant topical delivery and biotransformation on chemopreventive activity of apigenin in mouse skin. Dorsal and abdomen skin of female SENCAR mice were treated and compared with apigenin in different model vehicles: DMSO, acetone/DMSO (4:1) and propylene glycol/DMSO (4:1). Apigenin showed to be effective in the order of DMSO> acetone/DMSO > propylene glycol/DMSO. Although most of these vehicles may not be suitable for human use, one of the goals of the study was to investigate whether apigenin can get into the blood circulation through the skin. The results suggested that apigenin permeate only locally to the skin tissues, with no transdermal permeation observed. The biotransformation experiments concluded that sulfate and glucuronide metabolites are not involved in the chemopreventive activity of apigenin (Li, Pinch et al. 1996).

Incorporation of apigenin into a topical liposomal formulation showed success as substitutes for corticosteroid therapy in a clinical study. The apigenin-enriched, standardized chamomile extract were encapsulated into Natipide®II, a semi solid liposome gel, containing

20% of purified phospholipid fraction. The effectiveness of this formulation was compared to a non-liposomal oil-in-water cream, containing 5% Eumulgin® VL 75 as emulsifier. *In vitro* dissolution studies showed superior characteristics for liposomal creams. *In vivo* both were well tolerated and reduced inflammation, however, the liposomal formulation had slightly better therapeutic effect despite the same drug concentration. These results indicate that liposomes can enhance the penetration into the skin and the formulation strategy has a great importance in therapeutic efficiency. Based on this clinical study, apigenin-enriched formulations showed a great promise as a substitutes of corticosteroid therapy for e.g., eczema without severe side-effects (Arsic, Tadic et al. 2011).

Drug delivery of apigenin into the skin is not only as an alternative but as a first choice treatment showed promising results against skin cancer. This disease is becoming resistant to conventional chemotherapy therefore improved drug delivery like mitochondria and DNA targeting is needed. In a recent study the efficacy of apigenin-loaded PLGA nanoparticles for mitochondrial targeting was tested (Das, Das et al. 2013). PLGA (poly (lactic-co-glycolid acid) is a biodegradable, biocompatible and non-toxic material, one of the most successfully developed polymer. Its hydrolysis leads to two non-toxic metabolites: lactic acid and glycolic acid thus minimal toxicity is associated with the use of PLGA. It is approved by the FDA and European Medicine Agency for parenteral administration. Several formulations and methods of production for various drugs are well described. It is also able to protect the encapsulated drugs from degradation. Furthermore, sustained release and even specific organ or cell targeting can be achieved (Danhier, Ansorena et al. 2012). PLGA nanocarriers have the advantage of improving the permeability of the drug and showed enhanced delivery through mice skin. Thus, the aim of the study was to determine whether nano-encapsulated apigenin have greater anti-proliferative effect on UV-B and benzo(a)pyrene induced skin cancer in mice. Encapsulation was prepared using a one-step procedure of nanoprecipitation. Atomic force microscopy image showed that the majority of the particles are uniform and have spherical shape. The mean diameter was 101 nm with 87.2% encapsulation efficiency. Controlled release up to 72 h could be observed with a biphasic release profile characteristics, the initial burst release lasted for 16 hours. In order to determine the efficacy of nano-encapsulated apigenin histopathological sections and chromosomal aberrations were studied. Results showed that encapsulation improved the effectiveness due to small size and faster mobility; it reduced tissue damage and frequency of chromosomal aberrations. Moreover mitochondrial-apoptosis in cancerous tissues could be observed therefore nano-encapsulated apigenin appears to be promising in the skin cancer therapy (Das, Das et al. 2013). The same group found that PLGA nano-encapsulated apigenin had the ability to enter the nucleus thus directly target DNA. This suggests that PLGA nanoparticles can generate greater apoptotic effect and faster action and can prospectively be a new anti-cancer strategy in drug delivery to the skin (Das, Das et al. 2013).

Recently, a novel topical delivery system was developed for apigenin using ethosomes. This novel type of liposomes can enhance absorption and solubility due to 20-25 % (v/v) of lower alcohol content such as ethanol or propylene glycol. Their high deformability helps to transport the drug into the deeper layer of the skin more effectively than conventional liposomes. The optimal formulation was identified by uniform design experiment. Binary ethosomes were prepared to increase the stability and reduce aggregation using ethanol and propylene glycol. Lipoid S 75 (phosphatidylcholine containing 68-73% of soybean lecithin) was chosen as lipid phase. To compare the effectiveness of ethosomes, empty ethosomes with external apigenin, conventional liposomes and deformable liposomes (containing Tween 80) were prepared with mechanical dispersion method. All formulations were loaded with 0.02% (w/v) apigenin. The volume ratio of the lower alcohols significantly affected the mean ethosome size. With increasing amount of propylene glycol the size increases or decreases with the total amount of alcohols. The particle size of the formulations ranged between 36.61 ± 1.78 nm and 698.33 ± 124.30 nm, with zeta potentials between 10.14 ± 2.04 mV and 27.67 ± 3.23 mV. The optimum formulation had particle size distribution of 67.09 ± 4.10 nm with 19.30 ± 0.89 mV zeta potential. Entrapment efficiency of the optimum ethosomes was 91.22 ± 6.38 %, significantly higher than the other formulations 89.55 ± 1.57 % for liposomes and 81.93 ± 0.63 % for deformable liposomes. Increased level of Lipoid S 75 increased the entrapment efficiency of apigenin and the stability at room temperature for 30 days. Skin deposition and transdermal efflux experiments were also conducted. Skin deposition measurement revealed that increasing amount of propylene glycol decreased the skin deposition of apigenin, however, this increased with higher amount of Lipoid S 75 and lower alcohol concentration in the formulation. High concentration of lower alcohols can positively influence the flexibility and fluidity of the ethosomes therefore increased deformability can enhance the penetration of the drug through stratum corneum. Considering these findings, the optimum formulation should consists of 5% Lipoid S 75 and a mixture of lower alcohols in the ratio of 1:10 (v/v) propylene glycol to ethanol. Transdermal efflux experiments showed similar result regarding the increased propylene glycol ratio to ethanol. Ethanol is a one of the most used penetration enhancer and co-solvent but it often causes excessive transdermal drug flux. Propylene glycol was therefore added to increase the viscosity, hygroscopicity and stability thus enhancing the accumulation into deep skin layers. Superior skin targeting could be achieved *in vitro* with these optimized ethosomes. This improved permeability of apigenin into the skin is mostly related to the vesicle deformability and smaller particle size of ethosomes than conventional and deformable liposomes. Conventional liposomes did not penetrate deeply into the skin and in case of deformable liposomes the deposition of apigenin was even slower. However, there were number of differences between *in vivo* and *in vitro* skin deposition profiles, with ethosomes having the strongest effect on reducing inflammation induced by ultraviolet B (UVB) light. Ethosomes therefore can be a promising tool for superior topical delivery and targeting (Shen, Zhang et al. 2014).

Apigenin nanocrystals were prepared also as a skin protective formulation against UVB. Nanosuspension or so-called nanocrystals - when the drug particles are in crystalline state - are nanosized and carrier free colloidal dispersions consisting of drug particles with only a small amount of stabilizer e.g., Plantacare 2000 UP® (alkyl polyglycoside). The main advantages of nanocystals are increased dissolution velocity and saturation solubility which results in increased concentration gradient between the formulation and the skin and therefore prolonged contact with the skin due to their adhesiveness (Müller and Peters 1998). Nanocrystals can be produced with "top-down" (reduce particle size) or "bottom-up" (growth of particles from molecules) technologies. The "top-down" technology e.g., wet milling or high pressure homogenization is widely used in pharmaceutical research because of the easier scale up for the industry. Since 2001, smartCrystals® technology has been used. Nanocrystals prepared with this combination technology (pre-treatment step followed by high pressure homogenization step) show better characteristics e.g., physical stability. In this study the combination technology - bead milling and high pressure homogenization - was used to prepare apigenin smartCrystals. Significant reduction in size was observed after each bead milling process until the 5th passage therefore it was terminated after 7 cycles, followed by low pressure (300 bar) homogenization step. Generally low pressure (100-500 bar) yields smaller and more homogenous nanocrystals. The photon correlation spectroscopy (PCS) diameter of the prepared nanocrystals was 396 ± 12 nm with 0.205 PdI ± 0.007. Low PdI indicates better physical stability. Light microscopy measurement showed uniform crystal distribution and no large crystals (>1μm) or aggregates could be detected. 1% of Plantacare 2000 UP® stabilizer provide optimal stability. Zeta potential measurements indicated well charged surface stability (-38 mV). X-ray diffraction (XRD) revealed no amorphous stage which would reduce the stability and shelf life of the product. The developed smartCrystals can be easily utilized for further use as gel or cream where possibly a faster dissolution rate and increased permeation would be achieved. Furthermore, the *in vitro* antioxidant capacity of the nanosuspension is doubled compared to the original suspension (Al Shaal, Shegokar et al. 2011).

Oral Delivery

Orally administered apigenin has also a great potential in anticancer therapy but the extensive metabolism and poor aqueous solubility make it challenging. Nanocrystals can be further used to improve the oral bioavailability and also produce reproducible oral absorption. A new preparation method, SAS (supercritical antisolvent process) has been developed recently. This "bottom-up" process (recrystallization technology) requires no surfactant or other excipients and based on one kind of supercritical fluid technology. Briefly, the organic solution of the drug is injected into the supercritical fluid (generally CO_2) and the drug precipitates as fine particles as the solvent get extracted and the drug solution gets supersaturated.

The improved dissolution rate of the nanocrystals might be the key factor to increase the bioavailability and an effective formulation strategy for oral drug delivery (Zhang, Huang et al. 2013). Zhang et al. measured the physicochemical and pharmacokinetic properties of the coarse powder and the prepared apigenin nanocrystals by SAS process. Morphological evaluation of the nanocrystals by scanning electron microscopy revealed regular shape and smooth surface with particle size of 400-800 nm. The mean particle size was determined to be 562.5 ± 56 nm with 0.92 ± 0.21 PdI according to PCS measurements. The crystals were mainly unaltered considering XRPD (X-ray powder diffractometry) and DSC (differential scanning calorimetry) analysis which means beneficial physicochemical stability. Moreover, FT-IR analysis showed that apigenin remained chemically stable during the process. Conducting *in vitro* dissolution studies, nanocrystals have higher dissolution platform (more rapid and higher cumulative amount) which can be attributed to the enhanced saturated solubility due to the reduced particle size. Only 40% of the coarse powder was dissolved in a 120 min study under sink conditions (0.1 M PBS 6.8 with 0.5% Polysorbate). In contrast more than 90% dissolved within 20 min, as apigenin nanocrystals, demonstrating good *in vitro* dissolution behavior. The plasma concentration of apigenin following intravenous administration *in vivo* decreased rapidly with time and followed a biphasic pattern due to the initial distribution and metabolism in the tissues. The formulation administered orally resulted much lower serum concentration, but still higher than the coarse powder. The C_{max} and $AUC_{0-12\,h}$ were enhanced by 3.6 and 3.4 fold compared to apigenin coarse powder. The absolute bioavailability was enhanced from 2% up to 6.9%. It was concluded that nanocrystals with much smaller size and larger surface area can significantly improve the dissolution rate which contributes to higher C_{max} and $AUC_{0\text{-}12\ h}$ after oral administration. Furthermore, increased muco-adhesion and gastrointestinal transit time can be assumed.

In another paper mixed TPGS (D-α-tocopherol acid polyethylene glycol succinate) modified phospholipid micelles were designed to increase the oral bioavailability of apigenin. The vesicular structure of the phospholipid complexes have limited stability therefore TPGS - a water soluble derivate of natural Vitamin E - was added. It is formed by the esterification of Vitamin E succinate with polyethylene glycol (PEG) and therefore contains a hydrophilic head (PEG) and a lipophilic tail (Vitamin E) like other surface active molecules. The relatively low CMC (0.02% w/w) makes it an ideal molecular biomaterial for developing liposomes and nanoparticles. It has been used as a solubilizer, stabilizer and absorption enhancer. It can prolong the half-life of the drug in the plasma and enhance cellular uptake but most importantly inhibits P-glycoprotein increasing oral delivery (Zhang, Tan et al. 2012). The apigenin-phospholipid-TPGS micelles were prepared with thin film hydration method. The 3D molecular modeling showed that apigenin was positioned in the hydrophobic chains of lecithin, directed to the glycerol moiety and similarly in the hydrophilic PEG chains of TPGS where intramolecular hydrogen bonds and electrostatic forces possibly play a role in the interaction. The 3D structures is shown on Figure 4. The computer model prediction - that PEG chains are directed to the surface of the micelle – is confirmed by the negative zeta potential data (-12.94 mV). The complexation with phospholipids was verified by FT-IR and NMR spectroscopy measurements while the interaction with TPGS was studied with surface tensiometry. The encapsulation efficiency was 87.35%, drug loading was 12.6% with 137.1±3.4 nm particle size. *In vitro* release experiments showed an initial controlled release for 2 hours and the plateau was reached after 15 hours. Increased intestinal absorption up to 2.4 fold and higher cellular uptake with significant cytotoxicity effect on A549 cancer cell lines could be achieved.

In vivo 72.9% inhibition in S180 carcinoma mice were observed. These results suggest that TPGS micelles combined with phospholipid complex technology can be a novel way for oral drug delivery (Munyendo, Zhang et al. 2013).

Oral liquid dosage forms are therefore usually preferred; self-microemulsifying systems (SMEDDS) – a novel and promising technology for enhancing oral absorption has gained attention recently. SMEDDS contains oil, surfactant, co-surfactant and the drug. The drug can be solubilized in the oil phase and/or on the surface of the surfactant and co-surfactant in the microemulsion. Therefore the surfactant and co-surfactant play an important role in the solubilization of the drug. Mixing with water can produce thermodynamically stable oil-in-water microemulsion. The drug is dissolved in small droplets (size <100 nm) therefore the large surface area can enhance drug absorption (Pouton 2000). This system shows high drug entrapment and fast dissolution but the increase in the oral bioavailability is still needs to be estimated (Narang, Delmarre et al. 2007).

Zhao et al. investigated a self microemulsifying delivery system for apigenin, optimized by a simplex lattice experiment design. Ternary phase diagrams were also constructed from the excipients with high solubility of apigenin to obtain the most efficient concentration

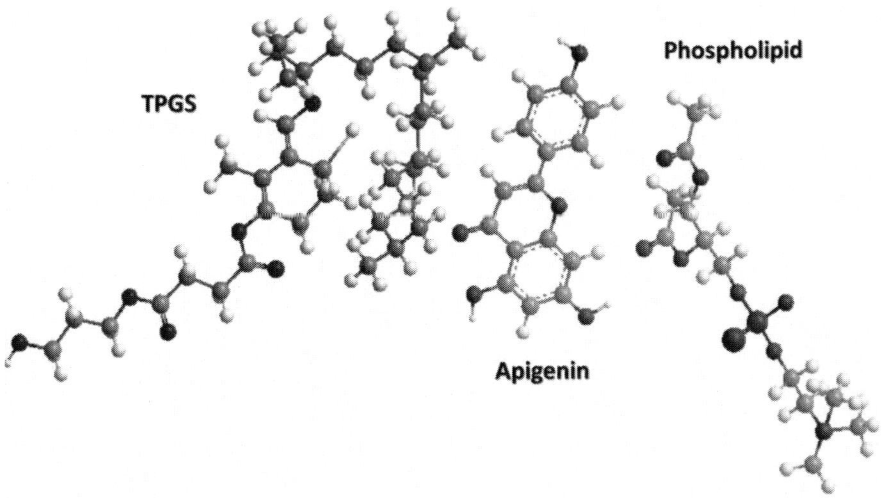

Figure 4. TPGS (D-α-tocopherol acid polyethylene glycol succinate) modified phospholipid micelles to increase the oral bioavailability of apigenin.

ranges for microemulsions. The apigenin exhibited the highest solubility in Capryol^{TM} 90 oil (1.39 ± 0.03 mg/mL), in Tween 80 surfactant (26.59 ± 1.16 mg/mL) and in Transcutol®HP co-surfactant (18.16 ± 0.43 mg/mL). The drug loading depends largely on drug solubility in the oil phase. The surfactant should be relatively hydrophobic (hydrophilic-lipophilic balance, HLB < 12) to obtain high self-microemulsifying ability (immediate formation of o/w droplets) and forms stable microemulsion when diluted with aqueous media (Kohli, Chopra et al. 2010). Although Tween 80 exhibited the highest apigenin solubility, it could not form SMEDDS with Transcutol®HP and Capryol^{TM} 90. PEG 400 was selected for the co-surfactant screening along with Transcutol®HP, however, its higher hydrophilic property could destroy the emulsion system. Finally, the optimal formulation consisted of 60% Cremophor®EL, 30% Transcutol®HP and 10% Capryol^{TM} 90. 7500-fold increase in water solubility of apigenin could be achieved with 17.1 nm average particle size and zeta potential -5.18 mV. The concentration of apigenin was selected to maximum 7 mg/g since at levels exceeding this apigenin precipitated during dilution with water. Transmission electron microscopy (TEM) images showed spherical droplets without agglomeration. 30 s was enough for the formation of microemulsion after dilution with distilled water and it did not precipitate after 8 hours (1 g of SMEDDS to 100 mL distilled water). The droplet size can be influenced by many factors and it plays an important role in the absorption. Interestingly, the size of this formulation was not affected by pH and ionic strength (diluted with distilled water, 0.9% NaCl, pH 6.8 PBS, 0.1 M HCl) and various mixing ways (vortex, magnetic stirring) but influenced by amount of apigenin. *In vitro* dissolution study demonstrated fast dissolution: 95 % of apigenin was released within 10 minutes. All of the results confirmed that SMEDDS could enhance the solubility of apigenin as a potential carrier for oral absorption (Zhao, Zhang et al. 2013).

Nanosized polymeric micelles with particle size ranging from 10-100 nm have been recognized as a potential drug delivery systems for anticancer drugs. Their small size allows accumulation in cancerous tissues and prolonged circulation time without being recognized by reticuloendothelial system (RES). Conventionally, biocompatible and self-assembly amphiphilic block copolymers can form core-shell structure in aqueous media. The hydrophobic fraction is able to incorporate poorly water soluble drugs as a core while the hydrophilic chains crosslinking outside and form a shell (Croy and Kwon 2006). In a recent study apigenin was successfully loaded into polymeric micelles composed of Pluronic P123 and Solutol HS15, to improve its water solubility with a thin-film dispersion method (Zhai, Guo et al. 2013). Pluronic P123 is one of the most widely used triblock copolymer with a structure PEO-PPO-PEO. The hydrophobic PPO group comprises 70%, thus providing suitable microenvironment for hydrophobic drug while 30% PEO can form a relatively thin shell with reduced stability, necessitating further modifications (Schillen, Jansson et al. 2008). Solutol HS15 (polyethylene glycol-660 hydroxystearate) is recorded in the European Pharmacopoeia as a non-ionic solubilizing agent containing 30% of free polyethylene glycol (PEG). In general, PEG used innanoformulations can increase stability and systemic circulation time (Vonarbourg, Passirani et al. 2006). The micelle formulations were optimized by central composite design. The solubility of apigenin was increased 148 times higher (320.8 μg/mL) than of crude apigenin (2.16 μg/mL) with low CMC concentration (4.23 x 10^{-5} mol/L) the following formulation: apigenin 1.68 mg, P123 92 mg and Solutol HS15 29 mg. The encapsulation efficiency and drug loading were 96.36% and 1.32 %. Under transmission electron microscope (TEM) the formulation had homogenous morphology. The size was 16.9 nm, smaller than blank micelles (18.9 nm), probably due to hydrogen bonds between phenolic hydroxyl group of apigenin and the carboxyl of the PEG chains. Moreover, *in vitro* drug dissolution study showed sustained release behavior and cytotoxicity studies on HepG2 and MCF-7 hepatoma cells demonstrated enhanced tumor inhibition (Zhai, Guo et al. 2013).

The solubility behavior of an active ingredient is the most challenging aspect, therefore there is a constant demand for new type of drug carriers in oral formulation development. Although solid dispersions (SDs) are known since 1961, their application is still increasing with new types of dispersing carriers e.g., silica nanopowder. In these dispersions the drug is incorporated in an inert carrier in solid state. Sugar was used as a carrier in the first generation SD. Second generation SD utilized amorphous polymeric carriers e.g., PEG in the late sixties, and more recently self-emulsifying and high surface active carriers belong to third generation SD. These carriers are able to improve dissolution rate significantly by reducing particle size, amorphous drug state and enhanced wettability and porosity can be achieved avoiding drug crystallization. Optimized manufacturing techniques make easier scale-up possible for industrial research (Chiou and Riegelman 1971, Leuner and Dressman 2000, Vasconcelos, Sarmento et al. 2007). One new type of dispersing carrier is carbon nanopowder (CNP). This nanomaterial is built up from carbon with diameter of less than 100 nm. It has many unique features like large surface area, chemical inertness and high dispersibility which helps to improve drug dispersion, and are thus widely used in targeted drug therapy. CNP-Apigenin was prepared by solvent evaporation. Briefly, apigenin and CNP were dissolved in ethanol and evaporated in rotary evaporator at 40°C until a clear powder mixture was formed. In comparison a homogenous physical mixture was prepared by grinding using a mortar and pestle. *In vitro* drug release and *in vivo* performance were also evaluated from the solid dispersions of apigenin prepared with CNP. All of the CNP-Apigenin had better dissolution profiles than apigenin powder alone (only 38% was dissolved). Approximately 92% of apigenin was released from the formulation with weight ratio 6:1 and the drug release dissolution profile was improved by 275% within 60 minutes compared to crude apigenin powder. DSC thermogram did not show the peak of apigenin suggesting it is molecularly dispersed and may be in amorphous form, which facilitated the dissolution. Scanning electron microscopy confirmed that apigenin is dispersed in CNP while XRD patterns show amorphous state. Oral bioavailability was tested in rats. The pharmacokinetic analysis showed 1.83 times increase in the $AUC_{0\text{-}t}$ for CNP-Apigenin indicating improved bioavailability. The CNP-Apigenin enhanced the relative oral bioavailability by 183%. The *in vivo* tests were in accordance with *in vitro* data suggesting that the poor bioavailability can be attributed to poor dissolution and reduced absorption. Preliminary intestinal toxicity test was carried out on jejunum mucosa of rats and did not show any degeneration, necrosis, edema or inflammation. Therefore, it can be assumed that CNP are safe and effective vehicles to enhance bioavailability for poor water soluble drugs. Moreover, CNP-Apigenin did not agglomerate and flowed freely, making it possible to formulate on a large scale. It was concluded that CNP is a promising SD carrier for clinical application (Ding, Zhang et al. 2014).

Dendrimers for Apigenin Delivery

Dendrimers are polymeric macromolecules with hyperbranched three dimensional structures (Figure 5). Controlled, globular structure and a large number of functionalities yet single molecular weight are significantly advantageous over traditional linear polymers. These attracted researcher's attention from the mid-1980s because of their unique shape and potential applications in drug delivery. Since then, a large number of dendrimers have been synthetized with various architectures. Two well documented synthetic strategies are used to prepare dendrimers, with the main difference being polymer growth (Hawker and Frechet 1990, Tomalia, Naylor et al. 1990). Basically, the polymer branches grow divergently or convergently start from a polyfunctional core (Liu and Fréchet 1999). A number of application are known for dendrimers. In the past decades biocompatible dendrimers were synthetized for drug delivery of e.g., vaccines and immunology. Recently, attempts have been made for cancer treatment and photodynamic therapy. Dendrimers can be useful tools as multifunction nanoparticulate systems for imaging, diagnostic and targeting in the therapy (Gillies and Fréchet 2005).

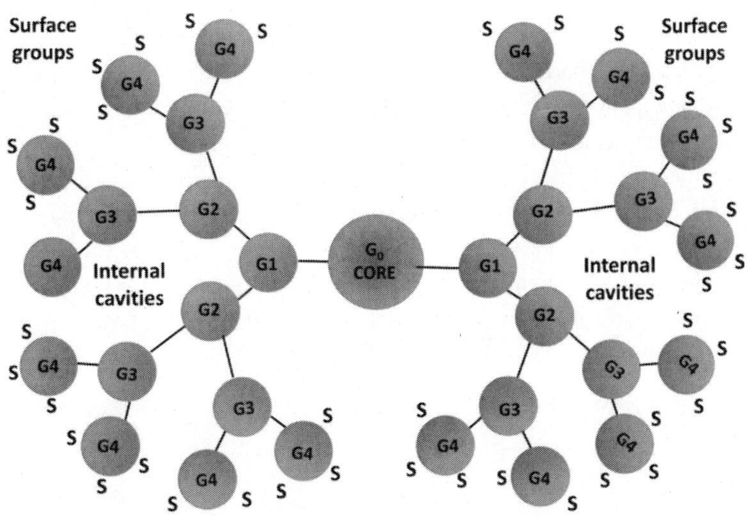

Figure 5. Internal cavities of the dendrimeric structure.

They are a technological breakthrough and their use in nanotechnology is progressively emerging. Although accurate selection of dendrimers is a key factor in drug delivery they have a great potential as an optimal drug delivery system for numerous drugs including DNA, proteins, gene therapy as well as diagnostic and solubilizing agent (Kesharwani, Jain et al. 2014). Apigenin was chosen as a flourophore core of benzylic dendrimers in a recent study. Fluorescent dendrimers have an advantage as analytical tools or organic light emitting devices (OLEDs). Apigenin possess weak blue light emission and three phenol groups therefore ideal for synthetizing dendrimer according to Fréchet method (Hawker and Frechet 1990, Liu and Fréchet 1999). Size and branching effects were studied with molecular dynamics simulation. Data suggest that larger asphericities can occur at the third and fourth generations of dendrimers. Aggregation phenomena can be suggested for nonspheric dendrimers from the fluorescent spectras (Vinš, Vermachová et al. 2013).

Further Prospects

Phytosome technology is a promising concept for enhancement of oral bioavailabilty. In water, phytosomes form liposome-like structure, however, there are significant differences (Kidd 2009, Semalty, Semalty et al. 2010). The phytosome is a molecular complex consisting of the drug and phosphatidylcoline which later is compatible with biological membranes. The phospholipids are amphiphilic molecules and easily incorporate polar or non-polar compounds, thus enhance the absorption through lipid-rich biological membranes. Since they can improve the bioavailability of both water soluble and insoluble materials, phytosomes may offer oral delivery for apigenin in the near future.

CONCLUSION

In the last few decades the use of nutraceuticals to prevent chronic illnesses like diabetes, cardiovascular diseases and cancer has been rapidly emerging. However, their low bioavailability *in vivo* limits their usage for medical therapy. Researchers are drawn towards nanotechnology which can change the pharmacokinetics and biodistribution thus improving the bioavailability and efficacy. Solubility improvement of apigenin is crucial in order to achieve therapeutic effect as discussed in the above section. The delivery of apigenin requires pharmaceutical formulation to provide protection of the therapeutically active molecular form until it reaches the target tissue. A few drug delivery approaches for apigenin have been developed in the past decades to address this issue, including novel techniques in the field of nanotechnology. The solubility behavior of an active ingredient – like apigenin – is still the most challenging part therefore there is a constant demand for new type of drug carriers in formulation development.

References

Abu-Yousif, A. O., Smith, K. A., Getsios, S., Green, K. J., Van Dross, R. T. and Pelling, J. C. (2008). Enhancement of UVB-induced apoptosis by apigenin in human keratinocytes and organotypic keratinocyte cultures. *Cancer Res.*, 68(8): 3057-3065.

Al Shaal, L., Shegokar R. and Müller R. H. (2011). Production and characterization of antioxidant apigenin nanocrystals as a novel UV skin protective formulation. *Int. J. Pharm.,* 420(1): 133-140.

Arsic, I., Tadic, V., Vlaovic, D., Homsek, I., Vesic, S., Isailovic G. and Vuleta, G. (2011). Preparation of novel apigenin-enriched, liposomal and non-liposomal, antiinflammatory topical formulations as substitutes for corticosteroid therapy. *Phytother. Res.,* 25(2): 228-233.

Asim, E. (1993). ß-Glucosidases, American Chemical Society.

Berrin, J. G., Czjzek, M., Kroon, P. A., McLauchlan, W. R., Puigserver, A., Williamson, G. and Juge, N. (2003). Substrate (aglycone) specificity of human cytosolic beta-glucosidase. *Biochem. J.,* 373(Pt 1): 41-48.

Birt, D. F., Hendrich, S. and Wang, W. (2001). Dietary agents in cancer prevention: flavonoids and isoflavonoids. *Pharmacology and Therapeutics,* 90(2-3): 157-177.

Chen, J., Lin, H. and Hu, M. (2003). Metabolism of Flavonoids via Enteric Recycling: Role of Intestinal Disposition. *Journal of Pharmacology and Experimental Therapeutics,* 304(3): 1228-1235.

Chiou, W. L. and Riegelman, S. (1971). Pharmaceutical applications of solid dispersion systems. *Journal of Pharmaceutical Sciences,* 60(9): 1281-1302.

Choi, J. S., Islam, M. N., Ali, M. Y., Kim, E. J., Kim, Y. M. and Jung, H. A. (2014). Effects of C-glycosylation on anti-diabetic, anti-Alzheimer's disease and anti-inflammatory potential of apigenin. *Food and Chemical Toxicology,* 64: 27-33.

Chunhua, L., Donglan, L., Xiuqiong, F., Lihua, Z., Qin, F., Yawei, L., Liang, Z., Ge, W., Linlin, J., Ping, Z., Kun, L. and S. Xuegang (2013). Apigenin up-regulates transgelin and inhibits invasion and migration of colorectal cancer through decreased phosphorylation of AKT. *J. Nutr. Biochem.,* 24(10): 1766-1775.

Croy, S. R. and Kwon, G. S. (2006). Polymeric micelles for drug delivery. *Curr. Pharm. Des.,* 12(36): 4669-4684.

Danhier, F., Ansorena, E., Silva, J. M., Coco, R., Le Breton, A. and Préat, V. (2012). PLGA-based nanoparticles: An overview of biomedical applications. *Journal of Controlled Release,* 161(2): 505-522.

Das, S., Das, J., Samadder, A., Paul, A. and Khuda-Bukhsh, A. R. (2013). Efficacy of PLGA-loaded apigenin nanoparticles in Benzo[a]pyrene and ultraviolet-B induced skin cancer of mice: mitochondria mediated apoptotic signalling cascades. *Food Chem. Toxicol.,* 62: 670-680.

Das, S., Das, J., Samadder, A., Paul, A. and Khuda-Bukhsh, A. R. (2013). Strategic formulation of apigenin-loaded PLGA nanoparticles for intracellular trafficking, DNA targeting and improved therapeutic effects in skin melanoma in vitro. *Toxicol. Lett.,* 223(2): 124-138.

Database (2014, May). USDA Database for the Flavonoid Content of Selected Foods, Release 3.1.

Ding, B., Chen, H., Wang, C., Zhai, Y., and Zhai, G. (2013). Preparation and in vitro evaluation of apigenin loaded lipid nanocapsules. *J. Nanosci. Nanotechnol.*, 13(10): 6546-6552.

Ding, S. M., Zhang, Z. H., Song, J., Cheng, X.D., Jiang, J. and Jia, X. B. (2014). Enhanced bioavailability of apigenin via preparation of a carbon nanopowder solid dispersion. *Int. J. Nanomedicine,* 9: 2327-2333.

Ferrari, M. (2005). Cancer nanotechnology: opportunities and challenges. *Nat. Rev. Cancer,* 5(3): 161-171.

Funakoshi-Tago, M., Nakamura, K., Tago, K., Mashino, T. and Kasahara, T. (2011). Anti-inflammatory activity of structurally related flavonoids, Apigenin, Luteolin and Fisetin. *Int. Immunopharmacol.*, 11(9): 1150-1159.

Gates, M. A., Vitonis, A. F., Tworoger, S. S., Rosner, B., Titus-Ernstoff, L., Hankinson, S. E. and Cramer, D. W. (2009). Flavonoid intake and ovarian cancer risk in a population-based case-control study. *Int. J. Cancer,* 124(8): 1918-1925.

Gillies, E. R. and Fréchet, J. M. J. (2005). Dendrimers and dendritic polymers in drug delivery. *Drug Discovery Today,* 10(1): 35-43.

Havsteen, B. H. (2002). The biochemistry and medical significance of the flavonoids. *Pharmacology and Therapeutics*, 96(2-3): 67-202.

Hawker, C. J. and Frechet, J. M. J. (1990). Preparation of polymers with controlled molecular architecture. A new convergent approach to dendritic macromolecules. *Journal of the American Chemical Society,* 112(21): 7638-7647.

Hollman, P. C., de Vries, J. H., van Leeuwen, S. D., Mengelers, M. J., Katan, M. B. (1996). Absorption of dietary quercetin glycosides and quercetin in healthy ileostomy volunteers. *American Journal of Clinical Nutrition*, 62: 1276-1282.

Horvathova, K., Novotny, L. and Vachalkova, A. (2003). The free radical scavenging activity of four flavonoids determined by the comet assay. *Neoplasma,* 50(4): 291-295.

Huang, Q., Yu, H. and Ru, Q. (2010). Bioavailability and delivery of nutraceuticals using nanotechnology. *J. Food Sci.,* 75(1): R50-57.

Huynh, N. T., Passirani, C., Saulnier, P., and Benoit, J. P. (2009). Lipid nanocapsules: A new platform for nanomedicine. *International Journal of Pharmaceutics,* 379(2): 201-209.

Kesharwani, P., Jain, K. and Jain, N. K. (2014). Dendrimer as nanocarrier for drug delivery. *Progress in Polymer Science,* 39(2): 268-307.

Kidd, P. M. (2009). Bioavailability and activity of phytosome complexes from botanical polyphenols: the silymarin, curcumin, green tea, and grape seed extracts. *Altern. Med. Rev.,* 14(3): 226-246.

Kim, H, Kim, H. W., Jung, S. (2008). Aqueous Solubility Enhancement of Some Flavones by Complexation with Cyclodextrins. *Bulletin of the Korean Chemical Society,* 29: 590-594.

Kohli, K., Chopra, S., Dhar, D., Arora, S. and Khar, R. K. (2010). Self-emulsifying drug delivery systems: an approach to enhance oral bioavailability. *Drug Discov. Today,* 15(21-22): 958-965.

Lefort, E. C. and Blay, J. (2013). Apigenin and its impact on gastrointestinal cancers. *Mol. Nutr. Food Res.,* 57(1): 126-144.

Leuner, C. and Dressman, J. (2000). Improving drug solubility for oral delivery using solid dispersions. *European Journal of Pharmaceutics and Biopharmaceutics,* 50(1): 47-60.

Li, B., Pinch, H. and Birt, D. F. (1996). Influence of vehicle, distant topical delivery, and biotransformation on the chemopreventive activity of apigenin, a plant flavonoid, in mouse skin. *Pharm. Res.,* 13(10): 1530-1534.

Li, B., Robinson, D. H. and Birt, D. F. (1997). Evaluation of properties of apigenin and [G-3H]apigenin and analytic method development. *Journal of Pharmaceutical Sciences,* 86(6): 721-725.

Liu, M. and Fréchet, J. M. J. (1999). Designing dendrimers for drug delivery. *Pharmaceutical Science and Technology Today,* 2(10): 393-401.

Loftsson, T. and Brewster, M. E. (1996). Pharmaceutical applications of cyclodextrins. 1. Drug solubilization and stabilization. *Journal of Pharmaceutical Sciences,* 85(10): 1017-1025.

Lugasi, A., Hóvári, J., Sági, K. V., Bíró, L. (2003). The role of antioxidant phytonutrients in the prevention of diseases. *Acta Biologica Szegediensis,* 47(1-4): 119-125.

Merisko-Liversidge, E. M. and Liversidge, G. G. (2008). Drug nanoparticles: formulating poorly water-soluble compounds. *Toxicol. Pathol.,* 36(1): 43-48.

Meyer, H., Bolarinwa, A., Wolfram, G. and Linseisen, J. (2006). Bioavailability of apigenin from apiin-rich parsley in humans. *Annals of Nutrition and Metabolism,* 50(3): 167-172.

Munyendo, W. L., Zhang, Z., Abbad, S., Waddad, A. Y., Lv, H., Baraza, L. D. and Zhou, J. (2013). Micelles of TPGS modified apigenin phospholipid complex for oral administration: preparation, in vitro and in vivo evaluation. *J. Biomed. Nanotechnol.,* 9(12): 2034-2047.

Müller, R. H. and Peters, K. (1998). Nanosuspensions for the formulation of poorly soluble drugs: I. Preparation by a size-reduction technique. *International Journal of Pharmaceutics,* 160(2): 229-237.

Nair, H. B., Sung, B., Yadav, V. R., Kannappan, R., Chaturvedi, M. M. and Aggarwal, B. B. (2010). Delivery of antiinflammatory nutraceuticals by nanoparticles for the prevention and treatment of cancer. *Biochem. Pharmacol.,* 80(12): 1833-1843.

Narang, A. S., Delmarre, D. and Gao, D. (2007). Stable drug encapsulation in micelles and microemulsions. *International Journal of Pharmaceutics,* 345(1–2): 9-25.

Nielsen, S. E., Young, J. F., Daneshvar, B., Lauridsen, S. T., Knuthsen, P., Sandstrom, B. and Dragsted, L. O. (1999). Effect of parsley (Petroselinum crispum) intake on urinary apigenin excretion, blood antioxidant enzymes and biomarkers for oxidative stress in human subjects. *Br. J. Nutr.,* 81(6): 447-455.

Nijveldt, R. J., van Nood, E., van Hoorn, D. E., Boelens, P. G., van Norren, K. and van Leeuwen, P. A. (2001). Flavonoids: a review of probable mechanisms of action and potential applications. *Am. J. Clin. Nutr.,* 74(4): 418-425.

Ollila, F., Halling, K., Vuorela, P., Vuorela, H. and Slotte, J. P. (2002). Characterization of Flavonoid–Biomembrane Interactions. *Archives of Biochemistry and Biophysics,* 399(1): 103-108.

Pápay, Zs. E. and Antal, I. (2014). Study on the antioxidant activity during the formulation of biological active ingredient. *European Scientific Journal Special Edition,* 3: 252-257.

Patel, D., Shukla, S. and Gupta, S. (2007). Apigenin and cancer chemoprevention: progress, potential and promise (review). *International Journal of Oncology,* 30(1): 233-245.

Pekić, B., Zeković, Z. and Lepojević, Z. (1994). Investigation of apigenin-7-O-β-glucoside hydrolysis by β-glucosidase from almonds. *Biotechnology Letters,* 16(3): 229-234.

Peterson, J. and Dwyer, J. (1998). Flavonoids: Dietary occurrence and biochemical activity. *Nutrition Research,* 18(12): 1995-2018.

Pinho, E., Grootveld, M., Soares, G. and Henriques, M. (2014). Cyclodextrins as encapsulation agents for plant bioactive compounds. *Carbohydrate Polymers,* 101(0): 121-135.

Pouton, C. W. (2000). Lipid formulations for oral administration of drugs: non-emulsifying, self-emulsifying and 'self-microemulsifying' drug delivery systems. *European Journal of Pharmaceutical Sciences*, 11, Supplement 2(0): S93-S98.

Prochazkova, D., Bousova, I. and Wilhelmova, N. (2011). Antioxidant and prooxidant properties of flavonoids. *Fitoterapia,* 82(4): 513-523.

Robert, H. G., Venkatakrishnan, G., George, W. F. and Dorothy, J. V. (1993). The Mammalian Cytosolic Broad-Specificity ?-Glucosidase. *β-Glucosidases, American Chemical Society,* 533: 83-112.

Ross, J. A. and Kasum, C. M. (2002). Dietary flavonoids: bioavailability, metabolic effects, and safety. *Annu. Rev. Nutr.*, 22: 19-34.

Saidman, E., Yurquina, A., Rudyk, R., Molina, M. A. A. and Ferretti, F. H. (2002). A theoretical and experimental study on the solubility, dissolution rate, structure and dipolar moment of flavone in ethanol. *Journal of Molecular Structure: THEOCHEM,* 585(1–3): 1-13.

Schillen, K., Jansson, J., Lof, D. and Costa, T. (2008). Mixed micelles of a PEO-PPO-PEO triblock copolymer (P123) and a nonionic surfactant (C12EO6) in water. a dynamic and static light scattering study. *J. Phys. Chem. B.,* 112(18): 5551-5562.

Semalty, A., Semalty, M., Rawat, M. S. and Franceschi, F. (2010). Supramolecular phospholipids-polyphenolics interactions: the PHYTOSOME strategy to improve the bioavailability of phytochemicals. *Fitoterapia,* 81(5): 306-314.

Shen, L. N., Zhang, Y. T., Wang, Q., Xu, L. and Feng, N. P. (2014). Enhanced in vitro and in vivo skin deposition of apigenin delivered using ethosomes. *Int. J. Pharm.*, 460(1-2): 280-288.

Shukla, S. and Gupta, S. (2010). Apigenin: a promising molecule for cancer prevention. *Pharm. Res.,* 27(6): 962-978.

Škerget, M., Kotnik, P., Hadolin, M., Hraš, A. R., Simonič, M. and Knez, Z. (2005). Phenols, proanthocyanidins, flavones and flavonols in some plant materials and their antioxidant activities. *Food Chemistry*, 89(2): 191-198.

Szente, L. and Szejtli, J. (1999). Highly soluble cyclodextrin derivatives: chemistry, properties, and trends in development. *Advanced Drug Delivery Reviews,* 36(1): 17-28.

Tomalia, D. A., Naylor, A. M. and Goddard, W. A. (1990). Starburst Dendrimers: Molecular-Level Control of Size, Shape, Surface Chemistry, Topology, and Flexibility from Atoms to Macroscopic Matter. *Angewandte Chemie International Edition in English,* 29(2): 138-175.

Tommasini, S., Raneri, D., Ficarra, R., Calabrò, M. L., Stancanelli, R. and Ficarra, P. (2004). Improvement in solubility and dissolution rate of flavonoids by complexation with β-cyclodextrin. *Journal of Pharmaceutical and Biomedical Analysis,* 35(2): 379-387.

Tong, X., Van Dross, R. T., Abu-Yousif, A., Morrison, A. R. and Pelling, J. C. (2007). Apigenin prevents UVB-induced cyclooxygenase 2 expression: coupled mRNA stabilization and translational inhibition. *Mol. Cell. Biol.,* 27(1): 283-296.

Vasconcelos, T., Sarmento, B. and Costa, P. (2007). Solid dispersions as strategy to improve oral bioavailability of poor water soluble drugs. *Drug Discovery Today,* 12(23–24): 1068-1075.

Vinš, P., Vermachová, M., Drašar, P., del Barrio, M., Jarne, C., Cebolla, V. L., de Cózar, A., Zangi, R. and Cossío, F. P. (2013). Size and branching effects on the fluorescence of benzylic dendrimers possessing one apigenin fluorophore at the core. *Tetrahedron,* 69(48): 10361-10368.

Vonarbourg, A., Passirani, C., Saulnier, P., Simard, P., Leroux, J. C. and Benoit, J. P. (2006). Evaluation of pegylated lipid nanocapsules versus complement system activation and macrophage uptake. *J. Biomed. Mater. Res. A.,* 78(3): 620-628.

Walle, T., Hsieh, F., DeLegge, M. H., Oatis, J. E., and Walle, U. K. (2004). High absorption but very low bioavailability of oral resveratrol in humans. *Drug Metab. Dispos.*, 32(12): 1377-1382.

Walle, U. K. and Walle, T. (2002). Induction of human UDP-glucuronosyltransferase UGT1A1 by flavonoids-structural requirements. *Drug Metab. Dispos.*, 30(5): 564-569.

Wang, W., Heideman, L., Chung, C. S., Pelling, J. C., Koehler, K. J. and Birt, D. F. (2000). Cell-cycle arrest at G2/M and growth inhibition by apigenin in human colon carcinoma cell lines. *Mol. Carcinog.*, 28(2): 102-110.

Wei, H., Tye, L., Bresnick, E. and Birt, D. F. (1990). Inhibitory effect of apigenin, a plant flavonoid, on epidermal ornithine decarboxylase and skin tumor promotion in mice. *Cancer Res.*, 50(3): 499-502.

Wu, D. G., Yu, P., Li, J. W., Jiang, P., Sun, J., Wang, H. Z., Zhang, L. D., Wen, M. B. and Bie, P. (2014). Apigenin potentiates the growth inhibitory effects by IKK-beta-mediated NF-kappaB activation in pancreatic cancer cells. *Toxicol. Lett.*, 224(1): 157-164.

Xiao, M., Shao, Y., Yan, W. and Zhang, Z. (2011). Measurement and correlation of solubilities of apigenin and apigenin 7-O-rhamnosylglucoside in seven solvents at different temperatures. *The Journal of Chemical Thermodynamics,* 43(3): 240-243.

Yin, F., Giuliano, A. E., Law, R. E. and Van Herle, A. J. (2001). Apigenin inhibits growth and induces G2/M arrest by modulating cyclin-CDK regulators and ERK MAP kinase activation in breast carcinoma cells. *Anticancer Res.*, 21(1a): 413-420.

Yuan, J. L., lv, Zh., Liu, Z. G., Hu, Z. and Zou, G. L. (2007). Study on interaction between apigenin and human serum albumin by spectroscopy and molecular modeling. *Journal of Photochemistry and Photobiology A: Chemistry,* 191(2–3): 104-113.

Zhai, Y., Guo, S., Liu, C., Yang, C., Dou, J., Li, L. and Zhai, G. (2013). Preparation and in vitro evaluation of apigenin-loaded polymeric micelles. *Colloids and Surfaces A: Physicochemical and Engineering Aspects,* 429(0): 24-30.

Zhang, J., Huang, Y., Liu, D., Gao, Y. and Qian, S. (2013). Preparation of apigenin nanocrystals using supercritical antisolvent process for dissolution and bioavailability enhancement. *Eur. J. Pharm. Sci.*, 48(4-5): 740-747.

Zhang, J., Liu, D., Huang, Y., Gao, Y. and Qian, S. (2012). Biopharmaceutics classification and intestinal absorption study of apigenin. *International Journal of Pharmaceutics,* 436(1-2): 311-317.

Zhang, Z., Tan, S. and Feng, S. S. (2012). Vitamin E TPGS as a molecular biomaterial for drug delivery. *Biomaterials,* 33(19): 4889-4906.

Zhao, L., Zhang, L., Meng, L., Wang, J. and Zhai, G. (2013). Design and evaluation of a self-microemulsifying drug delivery system for apigenin. *Drug Dev. Ind. Pharm.,* 39(5): 662-669.

In: Apigenin and Naringenin
Editor: Nilus M. Stacks
ISBN: 978-1-63463-987-3
© 2015 Nova Science Publishers, Inc.

Chapter 2

ANTIMETASTATIC POTENTIAL OF APIGENIN

Daniele Maggioni, PhD*
University of Milan-Bicocca, Department
of Surgery and Translational Medicine, Milan, Italy

ABSTRACT

Apigenin (4′, 5, 7-trihydroxyflavone) is a flavone, naturally occurring in many fruits and vegetables, such as oranges and grapefruits.

Apigenin has been progressively studied in the last decades for its potential use as anticancer agent, since its attractive anti oxidant, anti inflammatory properties associated to its low toxicity. In addition apigenin has been demonstrated to be non mutagenic.

However, although apigenin has proved to counteract tumour growth both in *in vivo* and *in vitro* studies, the mechanisms underlying remains largely unknown.

Apigenin exhibits a broad anti cancer effect, acting on protein kinases it modulates cell cycle and apoptosis, thus inhibiting cancer cell proliferation, moreover Apigenin might modulate carcinogens metabolism and disposition. Furthermore Apigenin counteracts tumour spreading, thorugh inhibition of cell motility and metastatization.

Metastasis is the final step of tumor progression and it represents one of the most challenging characteristics of malignant transformation and the main cause of cancer-related deaths. It is referred to as the diffusion of cancer cells outside the primary site of origin, through a complex process that encompasses cell adhesion, migration and immune escape.

Several studies have demonstrated apigenin ability in reducing cell motility, migration and invasion onto ontogenic different types of cancer. The molecular mechanisms underlying this effect are not completely understood, even if the modulation of MAPK and PI3K pathways are surely involved.

In this chapter the results of both *in vitro* and *in vivo* studies that evaluated the anti metastatic potential of apigenin will be summarized aiming at elucidate the molecular aspects involved. Moreover a comprehensive analysis of the methods and *in vivo* and *in vitro* models used in this field will be done.

---

* Corresponding author: Daniele Maggioni, phD. University of Milan-Bicocca, Department of Surgery and Translational Medicine, Milan, Italy. E-mail: daniele.maggioni1@unimib.it.

INTRODUCTION

Several epidemiological evidences have demonstrated that natural products in the diet might have anticancer properties. Because natural compounds generally have low toxicity and few side effects, they have been studied with the aim of potentially highlight new chemotherapeutic agents.

Among the natural compounds that researchers have focused on recently, flavonoids, polyphenolic compounds widely spread in largely consumed fruits and vegetables, have proved to be effective in cancer prevention. Furthermore scientific reports demonstrating that flavonoids are an effective natural inhibitor of cancer invasion are increasing.

Apigenin (4',5',7-trihydroxyflavone) belongs to the subfamily of flavones and it has been widely studied in the last decades because it is non toxic and non mutagenic. Apigenin is abundantly present in fruits, such as oranges and grapefruit and in plant-derived beverages like tea, chamomile and also in wine and beer. Several Apigenin biological effects depend on its anti oxidant and free radical scavenging ability, however Apigenin also exhibits anti inflammatory and cell cycle modulating capabilities. Moreover Apigenin seems to selectively affect cancer cell proliferation and vitality without affecting normal cell growth (Gupta et al., 2001; Maggioni et al., 2013).

Apart the great mass of scientific evidences supporting anti proliferative and apoptotic effect of Apigenin against cancer cells, some reports have also pointed out the Apigenin ability to interfere with the process leading to tumor spreading and cancer cell dissemination.

Although cancer trend rates are declining in Developed Countries, such as United States (Siegel et al., 2014), cancer still remains a global and steadily growing problem. According to the World Health Organization, deaths caused by cancer will double in the next 15 years. Therefore, it is mandatory to explore novel approaches for the treatment of cancer. In this field the evaluation of new strategies in cancer therapy is required, among them the exploitation of natural origin compounds has steadily grown in the last decade.

Metastasis represents one of the most serious and life hampering consequences of malignant tumor progression and it is responsible for over 90% of deaths from solid tumors (Wang et al., 2013).

Metastasis is the process which allows cancer cell to leave the site of primary lesion and establish a secondary tumor in a distant tissue. It is a multistep process, indeed it is often referred to as metastatic cascade, which begins with the cancer cell growth at the site of primary lesion, followed by the invasion of surrounding tissue, upon the degradation of basement membrane. During the acquisition of an invasive phenotype cancer cells lose their polarity and cell to cell adhesion, for epithelial cancer this process is referred to as Epithelial to Mesenchymal Transition (EMT) and it is considered a hallmark and the beginning of cancer metastasis. One of the first step in cancer invading bordering areas is the degradation of basal lamina (for epithelial cancer) and migration through the ExtraCellular Matrix (ECM).

CELL MIGRATION MODELS

Cell migration represents a topical issue in the cancer spreading and metastatization, thus this topic has been widely studied in both *in vitro* and *in vivo* experiments. Plenty of models have been used to study the different steps involved in this process: while *in vivo* model give a satisfactory consideration of all aspects involved, they have the disadvantage to be difficult to control and interpret, since any modification in one of the several aspects involved might result in a different outcome.

The study of cancer invasion should involve the analysis of cancer cell phenotype modification, interaction with ECM and microenvironment influences. It is conceivable that such a complex process is hardly reproducible in *in vitro* models, however *in vitro* model has the general advantages to allow a better experimental control, even if they lack the complexity of the in vivo environment. In addition 3D *in vitro* models could potentially recreate the same motility strategies used by cells in vivo, but they are quite complex and 2D approaches are generally preferred, although they have the intrinsic limitation to rely on amoeboid movements.

One of the earliest developed method to analyze cell migration *in vitro* is the Boyden Chamber migration assay where cells must pass through a micro porous filter which divides an upper from a lower space. The filter device may be embedded with ECM components allowing the evaluation of intrinsic cancer cell ability to degrade ECM. This model is effective in quantitatively evaluating the number of migrating cells, but it lacks information about migration mode and speed.

A more quantitative and informative model regarding the migration mode and speed is the Wound Healing assay, which measures the migration of cells along an artificial created scratch in a cell monolayer. Micrographs are taken at the beginning and after regular intervals of time and the migration of cells is quantified. However this method has the limitation to damage cells and matrix at the edge of the artificial created injury, in addition it requires a large number of cells (Wang et al., 2012).

Investigation assays on cell migration have shown significant advance over the past decade, many new methods that serve as complementation of scratch wound-healing assay have emerged to avoid matrix and cell damage. Self-assembled monolayers have been used to study cell to cell interaction during migration, moreover, by the use of microfluidic pump, it is possible to mimic the microenvironment during migration.

However this is an expensive and difficult method to set up. Apart intrinsic cell motility another fundamental aspect in cancer cell migration is the ability to degrade ECM by enzymatic activities: several analytic methods may provide good tools to evaluate such characteristics. Among them the evaluation of ECM degradation products by biochemical methods and the measurement of ECM degrading enzyme, such as Matrix MetalloProtease (MMP), quantity and activity are the most used.

Proteomics analysis of basement membrane and ECM surrounding cancer has allowed the *in vitro* recreation of ECM, in particular the use of ECM extract mixed with matrigel has significantly improved the reliability of *in vitro* artificially created ECM models to study cell migration (Benton et al., 2014).

Metastatization has been evaluated also in rodent models. These usually rely on dissemination of cells or the implantation of a tumor in a nude recipient and the presence of cancer cells in recipient tissues is analyzed after a period of time ranging between 2 to 4 weeks. An interesting model is based on chick chorioallantoic membrane invasion, where cancer cells are injected into a chick embryo and the ability of cancer cells to invade chorioallantoic membrane and to disseminate into chick fetus is evaluated.

ANTIMETASTATIC EFFECTS

As described above metastatic cascade comprises different steps that a cell must overcome to allow tumor spreading and to establish a secondary lesion outside the site of original cancer growth. In this part of the chapter we will address the available scientific data about the ability of Apigenin to prevent this process.

The first step a cell must overcome to begin the metastatic cascade is the entrance into the bloodstream, this relies on the modification of cell to cell attachment and the acquisition of motility. Indeed a cell has to gained such ability in order to move freely outside the cancer mass. Epithelial cells are characterized at the beginning by a high polarization and organization. In order to get motility a cancer cells must modify such an organization, therefore one of the first step in the metastatic cascade is the cytoskeletal reorganization. This process, addressed as the beginning of metastatization, is referred to as Epithelial to Mesenchymal Transition (EMT). Several molecules are involved in the regulation of cytoskeletal organizing, among them a key role is played by actin fibers, that are interconnected by catenins to transmembrane proteins, called Cadherins, that assure interconnection of cytoskeleton with ECM and surrounding cells.

Several reports have proved that flavonoids affect expression and organization of protein implied in cell to cell adhesion. Hu and colleagues (2008) demonstrated that Apigenin inhibited A2780 ovarian cancer cells migration *in vitro*. In particular Apigenin (20-40μM) reduced cell to ECM attachment, resulting in a more rounded appearance. Moreover Apigenin treated cells had a significant different organization of actin fiber, more pronounced around the periphery and not involved in stress fibers. All these modifications in actin and cytoskeletal organization were correlated to the reduction of Focal Adhesion Kinase (FAK) expression. FAK is a non receptor tyrosin kinase which may regulate cell adhesion by interacting with integrins. In particular FAK plays a pivotal role in focal adhesion promotion, multi-protein complexes needed for cells to interact with ECM. During the first steps of cell migration assembly and disassembly of focal complex occur. FAK overexpression has been correlated with tumor spreading. Apigenin treatment significantly attenuates FAK protein levels in A2780 cells. Although Apigenin minimally affected FAK activation (phosphorylation) and FAK mRNA levels, it significantly reduce FAK stability enhancing its proteasome degradation through ubiquitination and promoting FAK interaction with the proteasome.

Modulating cytoskeletal organization Apigenin potentially deeply affects cell motility and migration. Robust scientific evidences support this hypothesis, indeed the inhibition of cell motility by Apigenin has been reported by several *in vitro* experiments in different kinds of cancer cell lines. Apigenin (40-180μM for 48h) dose dependently reduced Huh-7 human hepatocellular carcinoma (HCC) cells migration in wound healing assay (Kim et al., 2011). This effect was confirmed also in human colorectal cancer cells and in human melanoma A518A2 cells (Chunhua et al., 2013; Sporlein et al., 2013). In addition Apigenin proved to be effective in reducing not only cell motility but also cell invasiveness through matrigel coated micro-pores (Frazen et al., 2009; Chunhua et al., 2013; Shi et al., 2013).

In a set of *in vitro* experiments indeed Frazen and coworkers (2009) demonstrated that Apigenin markedly inhibited cell migration and motility, evaluated by wound healing assay and by time lapse cell spreading. In particular Apigenin effect seemed to depend on the reduction of actin stress fibers. Actin polymerization coupled with myosin filament contractions and with integrin interactions are the leading forces for cell motility. Stress fibers are actin polymer connected with focal complex required for contractive tension and therefore required for transient cell attachment and motility. In the focal complex both stress fibers and integrins, transmembrane proteins that link actin cytoskeleton to ECM, join together to drive transient contraction needed for cell migration. In PC-3 cells Apigenin reduced cell substrate attachment ability and the formation of migration structure such as invadopodia. The molecular mechanisms underlying Apigenin inhibition of cytoskeletal remodeling seems to depend on the inactivation of FAK, whose activity is required for stress fibers and focal adhesion formation (Frazen et al., 2009). Apigenin interference with actin fibers organization and inhibition of β-catenin signaling has been highlighted also in melanoma cell lines (Sporlein et al., 2013). The inhibition of Catenins-actin interactions has a strong effect on cell motility, since it prevents cell detachment and migration.

Wang and coworkers (2013) demonstrated that Apigenin prevented CXCR4 expression in arsenic-transformed human epithelial bronchial Beas-2B cells. CXCR4 is one of the most studied chemokine receptor and its expression has been correlated with metastatization and poor survival. Apigenin-induced down-regulation of CXCR4 was not related to proteolytic degradation, but mainly due to the down-regulation of nuclear factor κB (NF-κB) transcriptional activity.

Upon detachment from the mass of cancer a cell must invade surrounding tissue in order to metastatize, therefore acquisition of ECM degrading capability are needed. At the beginning of these process proteolytic enzymes are recruited to cell surface areas characterized by high density of integrins and adhesion receptors to degrade basement membrane and ECM. Among the enzymes with ECM remodeling and degrading ability, both membrane associated and extracellular released MMPs play a pivotal role in facilitating cell migration through ECM. In particular MMP2 and MMP9, also referred to as gelatinases for their ability to digest collagen IV, the major component of basement membrane, are often overexpressed by cancer cells.

MMP activity is required in several physiological process and it is finely regulated both at the activation and at the inhibition process. Indeed MMPs are secreted as inactive enzymes and the proteolytic removal of the pro domain is required for their activation. This process is regulated by enzymes such as Urokinase Plasminogen Activator (uPA) or other MMPs, such as MMP14.

Inhibitors of MMPs activation and activity are present in the tissue, among them a fundamental role in MMP regulation is played by Tissue Inhibitor of MetalloProtease (TIMP).

Due to the crucial role played by MMP in cancer spreading and metastatization, several studies have been performed to evaluate the potential beneficial effect of plant derived flavonoids able to modulate MMP expression and activity. Few studies have also addressed the ability of Apigenin to inhibit ECM degrading enzymes. Noh et al. (2010) analyzed the Apigenin effect on the MMP mediated invasiveness of different human metastatic cell lines: small intestine cancer derived Caski, breast cancer derived MDA-231 and hepatocellular carcinoma cells SK-Hep1. They reported that Apigenin, at doses as low as 2,5μM, significantly prevented cell migration *in vitro* by the down-regulation of MMP9 expression. Apigenin effect was noted not only in standard conditions but also in the presence of phorbol 12-myristate 13-acetate (PMA), known to induce MMP9 expression in these cell lines.

Apigenin inhibition of PMA induced MMP9 expression was reported also by Lindenmeyer and colleagues (2001), that highlighted that Apigenin prevention of MDA-231 cell migration and matrigel invasion depended on the uPA activator expression reduction.

Sporlein et al., (2013) investigated the anti-cancer effect of Apigenin-copper complex and reported that it inhibited wound healing velocity of 518A2 melanoma cells. The mechanism underlying this inhibition was the reduction of MMP2 and MMP9 expression, secretion and activity evaluated by western blot and zymography.

Chunhua et al., (2013) demonstrated the Apigenin inhibition of cell motility in three kinds of colorectal adenocarcinoma cell lines SW480, DLD-1, LS174T. Apigenin dose dependently increased both mitochondrial and nuclear expression of Transgelin, a 22kDa protein, mainly expressed in smooth muscle rich tissues, where it acts binding actin in stress fibers. However Transgelin expression has been demonstrated also outside muscle and this protein is involved in cell migration during malignancies. Indeed it acts as actin fiber organizers in podosomes and invadopodia formation (Assinder et al., 2009), moreover it has been reported the Transgelin ability to inhibit MMP9 activity (Nair et al., 2006). In Chunhua's study the Apigenin induced increase of Transgelin expression was related to AKT phosphorylation. Therefore it is conceivable that Apigenin reduced MMP9 activity by promoting Transgelin expression through AKT activation.

Although Apigenin effect might vary according to cell and cancer type, these data confirme its ability to act as a suppressor of MMP and especially MMP9 expression.

When a cancer cell has gained freedom from surrounding cells and it has moved through the nearby tissue thanks to ECM degrading enzyme activity, it has to get into the blood or into the lymphatic flood. This requires the trans-endothelial migration of cancer cells known as intravasation and the survival in the circulatory system. The blood system has long been considered the main route for cancer metastatic spread, but increasing evidences highlighted the lymphatic system importance in cancer cell dissemination. Indeed lymph vessels are more easily invaded by cancer cells since they lack basement membrane and tight junctions between endothelial cells (Zijl et al., 2011). Few reports have evaluated Apigenin effect on this complex phenomenon. Lee and coworkers (2008) demonstrated that Apigenin significantly reduced intravasation ability of cancer cells. In a smart *in vitro* assay MDA-MB-231 human breast carcinoma cells were plated on a monolayer of endothelial Huvec cells and stimulated with HGF for 30', thereafter cancer cells to endothelial cells interaction were evaluated by immunofluorescence. When cancer cells were pretreated for 2h with 40μM Apigenin the number of interactions was markedly reduced.

Moreover Apigenin also reduced *in vivo* intravasating cells. Indeed, a model based on chick chorioallantoic membrane invasion was used *in vivo* to evaluate the number of intravasating and colonizing cells. In this experiments HGF was used as metastatization promoter and 2 hours Apigenin pretreatment proved to significantly reduce the incidence of lower chorioallantoic membrane colonization and the occurrence of lung and stomach metastasis (Lee et al., 2008). HGF promotes cell invasion by increasing integrins avidity for specific ligands, therefore promoting cell to matrix substrate adhesion.

Further experiments demonstrated that Apigenin intravasation inhibition depends on the modulation of β4 integrin avidity mediated by PI3K, since Apigenin effect was reverted by anti β4 integrin blocking antibodies and by PI3K specific inhibitors such as wortmannin.

Even if a cancer cell managed to successfully implant in a distant site, it must gained the ability to growth in the site of secondary lesion. Therefore the ability to obtain energy supply is mandatory, neo-angiogenesis is a key step for a tumor to be supplemented of oxygen and nutrition. Several growth factor, referred to as angiogenic activators, might induce new blood vessels formation and are secreted by invasive tumor cells (Weng et al., 2012).

An anti angiogenic effect of Apigenin has been demonstrated in hepatocellular carcinoma cell Huh-7, since Apigenin dose dependently reduced the levels of secreted Vascular Endothelial Growth Factor (VEGF), a key regulator of angiogenesis, and of MMP8, a protease mainly involved in ECM remodeling during new capillaries sprouting (Kim et al., 2012).

The inhibitory effect of Apigenin on VEGF expression was confirmed in a study by Mirzoeva et al. (2014) on prostate carcinoma cell line PC-3 and LNCaP. In details Apigenin inhibited TGF-β1 induced VEGF expression, through negative modulation of Smad2 and Smad3 activation and nuclear translocation. Moreover Apigenin also reduced activation of SRC and AKT kinases and their constitutively active forms reversed Apigenin induced VEGF down-regulation.

Apigenin anti angiogenic activity does not affect only cancer cells, on the contrary it was demonstrated also the inhibition of the spontaneous tube formation *in vitro* by Huvec embryonal endothelial cells (Kim et al., 2003), thus sustaining the broad anti angiogenic Apigenin effect.

Mechanisms underlying Apigenin inhibition of angiogenesis were clarified in a work by Liu and coworkers (2006) dealing with the anti-metastatic effect of Apigenin on A549 lung carcinoma cells. In this study Apigenin doses over 20μM were effective in inhibiting VEGF transcriptional activation through the reduction of the levels of the hypoxia-inducible factor 1α (HIF-1α) expression in the cells. Indeed HIF-1α is known as one of the transcription factors mainly involved in the regulation of VEGF promoter. These data were confirmed by experiments performed by Fang and coworkers (2006), that reported the Apigenin induced inhibition of HIF-1a and VEGF expression in several cancer cell lines, including the human prostate cancer cells, PC-3, DU145 and LNCaP; human ovarian cancer cells, OVCAR-3, and human colon cancer cells, HCT-8 and the human breast carcinoma cell lines MCF-7.

Osada and coworkers (2004) clarified the mechanism of Apigenin inhibition of angiogenesis, indeed they demonstrated that Apigenin reduced HIF-1a stability, modulating its interaction with the Heat Shock Protein 90, in Huvec endothelial cells.

Apigenin anti migratory and anti metastatization activity has been demonstrated also in *in vivo* models. Apigenin inhibition of B16-BL6 melanoma cells metastatic potential was demonstrated in a mice model, where Apigenin somministration significantly reduced the number of lung colonies. Furthermore 25 and 50mg/Kg Apigenin was more effective than the same doses of taxoxifen (Caltagirone et al., 2000). In a similar contest Lee and coworkers (2008) demonstrated the Apigenin inhibition of breast cancer MDA-231 cells metastatization. In the same study Apigenin also proved to be effective in preventing HGF breast cancer cells colonization to lung and liver in a model based on chick embryo metastatization. The ability of tumor cells to intravasate into the circulation followed by spontaneous metastasis to the lung or liver was assessed. The same anti-metastatic potential of Apigenin was reported by Chunhua et al., (2013) for SW480 colon carcinoma cells against intestine and liver metastasis in nude mice.

Apart reports in mouse model, only an additional report exists about the anti-metastatic potential of Apigenin in a rat model, indeed low Apigenin dose (1.5mg/kg) administration reduced the incidence of lymphatic vessels invasion and cancer metastasis of adenocarcinoma cells induced by azoxymethane (Tatsuta et al., 2000).

CONCLUSION

Data collected in this chapter strongly sustained the hypothesis of the potential use of Apigenin as anti-metastatic agent. Indeed Apigenin proved to be effective both *in vitro* and *in vivo* in preventing cancer spreading and dissemination of cancer cells to distant tissues.

Apigenin acting on different intracellular signal transduction pathways such as MAPK and PI3K, may modulate expression and activities of several key players of cell migration. Interestingly in fact, Apigenin has been shown to affect almost all the steps involved in the metastatic cascade, from cytoskeleton organization to cell to cell adhesion, ECM degradation and angiogenesis.

Moreover the safe, non-toxic and non-mutagenic profile of Apigenin makes it a very attractive candidate for chemotherapy.

Furthermore Apigenin might be exploited thanks to its broad anticancer effects, since it proved to selectively affect cancer cell growth, proliferation, migration and dissemination.

One main drawback remains to be overcome for Apigenin, which is its poor availability if orally assumed, therefore finding of new ways of administration or the use of new technologies to improve the bioavailability of Apigenin is mandatory.

REFERENCES

Assinder, S. J., Stanton, J. A., Prasad, P. D. Transgelin: an actin-binding protein and tumour suppressor. *Int. J. Biochem. Cell Biol.* 2009 Mar.;41 (3):482-6.

Benton, G., Arnaoutova, I., George, J., Kleinman, H. K., Koblinski, J. Matrigel: From discovery and ECM mimicry to assays and models for cancer research. *Adv. Drug Deliv. Rev*. 2014 Jul. 2. pii: S0169-409X(14) 00137-9.

Caltagirone, S., Rossi, C., Poggi, A., Ranelletti, F. O., Natali, P. G., Brunetti, M., Aiello, F. B., Piantelli, M. Flavonoids apigenin and quercetin inhibit melanoma growth and metastatic potential. *Int. J. Cancer*. 2000 Aug. 15; 87(4):595-600.

Chunhua, L., Donglan, L., Xiuqiong, F., Lihua, Z., Qin, F., Yawei, L., Liang, Z., Ge, W., Linlin, J., Ping, Z., Kun, L., Xuegang, S. Apigenin up-regulates transgelin and inhibits invasion and migration of colorectal cancer through decreased phosphorylation of AKT. *J. Nutr. Biochem.* 2013. 24(10):1766-75.

Czyz, J., Madeja, Z., Irmer, U., Korohoda, W., Hülser, D. F. Flavonoid apigenin inhibits motility and invasiveness of carcinoma cells in vitro. *Int. J. Cancer.* 2005 Mar. 10;114(1):12-8.

Franzen, C. A., Amargo, E., Todorović, V., Desai, B. V., Huda, S., Mirzoeva, S., Chiu, K., Grzybowski, B. A., Chew, T. L., Green, K. J., Pelling, J. C. The chemopreventive bioflavonoid apigenin inhibits prostate cancer cell motility through the focal adhesion kinase/Src signaling mechanism. *Cancer Prev. Res*. (Phila.). 2009. 2(9):830-41.

Gupta, S., Afaq, F., Mukhtar, H. Selective growth-inhibitory, cellcycle deregulatory and apoptotic response of apigenin in normal versus human prostate carcinoma cells. *Biochem. Biophys. Res. Commun.* 2000;287: 914–20.

Hu, X. W., Meng, D., Fang, J. Apigenin inhibited migration and invasion of human ovarian cancer A2780 cells through focal adhesion kinase. *Carcinogenesis.* 2008. 29(12):2369-76.

Kim, B. R., Jeon, Y. K., Nam, M. J. A mechanism of apigenin-induced apoptosis is potentially related to anti-angiogenesis and anti-migration in human hepatocellular carcinoma cells. *Food Chem. Toxicol.* 2011. 49 (7):1626-32.

Maggioni, D., Garavello, W., Rigolio, R., Pignataro, L., Gaini, R., Nicolini, G. Apigenin impairs oral squamous cell carcinoma growth in vitro inducing cell cycle arrest and apoptosis. *Int. J. Oncol.* 2013;43(5):1675-82.

Mirzoeva, S., Franzen, C. A., Pelling, J. C. Apigenin inhibits TGF-β-induced VEGF expression in human prostate carcinoma cells via a Smad2/3- and Src-dependent mechanism. *Mol. Carcinog.* 2014 Aug.;53(8):598-609.

Nair, J. Solway, D. D. Boyd. Expression cloning identifies transgelin (SM22) as a novel repressor of 92-kDa type IV collagenase (MMP-9) expression. *J. Biol. Chem.*, 281 (2006), pp. 26424–26436.

Noh, H. J., Sung, E. G., Kim, J. Y., Lee, T. J., Song, I. H. Suppression of phorbol-12-myristate-13-acetate-induced tumor cell invasion by apigenin via the inhibition of p38 mitogen-activated protein kinase-dependent matrix metalloproteinase-9 expression. *Oncol. Rep.* 2010. 24(1):277-83.

Osada, M., Imaoka, S., Funae, Y. Apigenin suppresses the expression of VEGF, an important factor for angiogenesis, in endothelial cells via degradation of HIF-1alpha protein. *FEBS Lett.* 2004 Sep. 24;575(1-3):59-63.

Shi, M. D., Liao, Y. C., Shih, Y. W., Tsai, L. Y. Nobiletin attenuates metastasis via both ERK and PI3K/Akt pathways in HGF-treated liver cancer HepG2 cells. *Phytomedicine.* 2013 Jun. 15;20(8-9):743-52.

Siegel, R., Naishadham, D., Jemal, A. Cancer Statistics, 2012. *CA Cancer J. Clin.* 2012 Jan.-Feb.;62(1):10-29.

Spoerlein, C., Mahal, K., Schmidt, H., Schobert, R. Effects of chrysin, apigenin, genistein and their homoleptic copper(II) complexes on the growth and metastatic potential of cancer cells. *J. Inorg. Biochem.* 2013. 127:107-15.

Tatsuta, A., Iishi, H., Baba, M., Yano, H., Murata, K., Mukai, M., Akedo, H. Suppression by apigenin of peritoneal metastasis of intestinal adenocarcinomas induced by azoxymethane in Wistar rats. *Clin. Exp. Metastasis.* 2000;18(8):657-62.

Wang, Y., Chen, Z., Xiao, L., Du, Z., Han, X., Yu, X., Lu, Y. Evaluating cell migration in vitro by the method based on cell patterning within microfluidic channels. *Electrophoresis.* 2012. 33(5):773-9.

Wang, L., Kuang, L., Hitron, J. A., Son, Y. O., Wang, X., Budhraja, A., Lee, J. C., Pratheeshkumar, P., Chen, G., Zhang, Z., Luo, J., Shi, X. Apigenin suppresses migration and invasion of transformed cells through down-regulation of C-X-C chemokine receptor 4 expression. *Toxicol. Appl. Pharmacol.* 2013. 272(1):108-16.

Weng, C. J., Yen, G. C. Flavonoids, a ubiquitous dietary phenolic subclass, exert extensive in vitro anti-invasive and in vivo anti-metastatic activities. *Cancer Metastasis Rev.* 2012 Jun.;31(1-2):323-51.

Zhu, Y., Mao, Y., Chen, H., Lin, Y., Hu, Z., Wu, J., Xu, X., Xu, X., Qin, J., Xie, L. Apigenin promotes apoptosis, inhibits invasion and induces cell cycle arrest of T24 human bladder cancer cells. *Cancer Cell Int.* 2013 Jun. 1;13(1):54.

Zijl, F., Krupitza, G., Mikulits, W. Initial steps of metastasis: cell invasion and endothelial transmigration. *Mutat. Res.* 2011 Jul.-Oct.;728(1-2):23-34.

In: Apigenin and Naringenin
Editor: Nilus M. Stacks

ISBN: 978-1-63463-987-3
© 2015 Nova Science Publishers, Inc.

Chapter 3

NANOTECHNOLOGY BASED APIGENIN DRUG DELIVERY SYSTEMS

Ranjita Shegokar*
Freie Universität Berlin, Institute of Pharmacy, Department Pharmaceutics, Biopharmaceutics & NutriCosmetics, Berlin, Germany

ABSTRACT

Few last decades witnessed increasing research reports on flavonoids because of their wide range of applications. Flavonoids have been used in completely different fields ranging from the food industry to medicine. Apigenin is one of the several flavonoids being researched and reported to have relatively low toxicity compared to other flavonoids. The main limitation in exploration of apigenin's effects is mainly due to the low water solubility which causes many formulation issues and hinders bioavailability of compound. Successful attempts are made to incorporate apigenin in various nanoformulations like micelles, polymeric nanoparticles and lipid nanocapsules. Several fold solubility enhancement of apigenin is mainly observed when nanocrystal technology is used. This chapter takes extensive overview of nanotechnology based apigenin encapsulated drug delivery systems and their performance *in vitro* and *in vivo*.

INTRODUCTION

In last few decades flavonoids have attracted a growing interest within the scientific community because of their wide range of applications. They have been used in different fields ranging from food industry, medicine, and cosmetics. Health and food industry have focused their interest in the large spectrum of biological and pharmaceutical activities of flavonoids. Apigenin, (4′, 5, 7-trihydroxyflavone) is one of the several active ingredients found naturally in many fruits and vegetables (Figure 1) and being studied for various medicinal activities. Apigenin is bioflavonoid and widely found in citrus fruit and in several

* E-mail: ranjita@arcsindia.com.

popular spices, including basil, oregano, tarragon, cilantro and parsley. It has anti-inflammatory effects, [1] free radical scavenging [2] and antiphlogistic, spasmolytic [3], analgesic [4] and antihistamine ones.

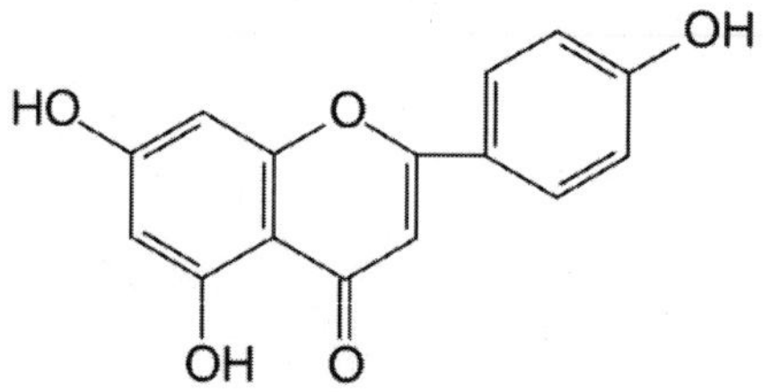

Figure 1. Structure of apigenin

Apigenin has been shown to possess growth inhibitory properties in several cancer lines including breast, colon, skin, thyroid leukemia cells and pancreatic, tumor inhibition and enzyme inhibitory properties. The extensive bioactivities such as cardiovascular effects, lowering blood pressure are also exerted by apigenin. It is also reported to be non mutagenic and of low toxicity compared to other related flavonoids [5].

Due to the poor solubility (both in aqueous and organic media) and resulting slow dissolution velocity apigenin have bioavailability problems [6-8]. Poorly soluble actives beside their lipophilic nature have good permeability through cell membrane. The few dissolved molecules can permeate a membrane with low dissolution from the crystal of the active showing very poor drug concentration gradient available for absorption. Thus, a low dissolution velocity is the rate limiting step. According to Biopharmaceutical classification (BCS) system, these drugs are classified under BCS-II, having poor solubility and high permeability. Apigenin is categorized as BCS class II drug. It is absorbed in the entire intestine with the main absorption site at duodenum by passive and active carrier mediated transport mechanism. In both transports, apigenin permeability is concentration dependent. Successful attempts are made to incorporate apigenin in micelles, polymeric nanoparticles and lipid nanocapsules. Several fold solubility enhancement of apigenin is mainly observed when nanocrystal technology is used. This chapter focuses on the formulation advances in apigenin drug delivery research and there *in vitro* and *in vivo* effects. The potential and sustainability of these technologies is also discussed briefly.

The solubility is important parameter of active compound and is affected by several parameters like temperature, size, and solvent. Min et al. studied the effect of temperature on solubility of apigenin and apigenin-7-O-rhamnosylglucoside in alcoholic solvents like methanol, ethanol, 1-propanol, 1-butanol and some other solvents like acetone, and ethyl acetate. Results indicated that the temperature had profound effect on solubility of apigenin [9].

POLYMERIC NANOPARTICLES

Polymeric nanoparticles are most widely studied nanoparticles. The preparation methods include solvent evaporation, salting-out, dialysis, supercritical fluid technology, micro-emulsion, and interfacial polymerization. Several products based on polymeric nanoparticles are accepted by Food and drug Administration (FDA). In recent years, there are number of synthetic drugs being explored and their nanoformulations are studied [10]. However, the interest to treat disease using natural active or remedies is booming.

Apigenin encapsulated PLGA (Polylactic-co-glycolide) nanoparticles of about 100 nm (zeta potential = -12 mV) prepared by solvent displacement showed improved anticancer potential in cultured Human skin melanoma cell line A375 and skin keratinocyte cell line HaCaT *in vitro. In vivo* intracellular ROS (reactive oxygen species), depolarization of mitochondrial membrane potential, ATPase enzyme activity were carried out in Swiss Albino mice. Significantly higher rate of intracellular ROS accumulation in animals treated with nanoparticles thereby depleting levels of GSH (glutathione) and SOD (superoxide dismutase) levels are observed. ROS generated by nanoparticle accumulation caused physiological changes in mitochondria as detected by changed transmembrane potential which was result of translocation of Bax protein to mitochondria. The data showed that the encapsulated APIGENIN in polymeric nanoparticles facilitated the intracellular transport, ROS generation and targeted DNA [11].

To treat skin cancer, PLGA apigenin nanoparticles were developed and tested in ultra-violet B and Benzo(a) pyrene induced skin tumor. Activity of nanoparticles to dysfunction mitochondria was tested in mice. Markers for mitochondrial physiology (cytochrome c (cyt c) release, mRNA and protein expressions of Apaf-1, bax, bcl-2, cyt c, cleaved caspase-9 and 3), chromosomal aberrations, ROS accumulation and mitochondrial transmembrane potential were determined. Histopathology of dorsal skin and blood plasma analysis performed after oral and topical treatment thrice a week. The administration of nanoparticles encapsulated with apigenin reduced the rate of tumor formation and the percent of chromosomal aberrations formation. Down regulation of PCNA (Proliferating cell nuclear antigen) and activation of caspase-3 in tumor cells in mice confirmed superior apoptotic activity of apigenin nanoparticles compared to coarse apigenin [12].

In another study, thin film dispersion method was used to prepare polymeric micelles (~17 nm) of Pluronic® P123 and Solutol® HS 15 and formed micelles evaluated in HepG2 and MCF-7 cancer cell showed potential for improvement of apigenin solubility [13]. Mixed micelles of apigenin stabilized with TPGS (water soluble Vit E) are reported to improve oral bioavailability and could target cancer [14]

NANOSUSPENSIONS

Several approaches are used to overcome solubility problem associated with these drugs, one of them is production of nanocrystals which not only increases drug surface area but also increases the dissolution velocity, which ultimately exhibit an increased saturation solubility Cs [15-19]. Enhanced saturation solubility (Cs) further leads to an increased concentration gradient at membranes promoting better permeation. Literature reports confirm improved performance when used orally [20-22] in the form of nanosuspensions or drug nanocrystals. The administration of nanosuspension via systemic administration via intravenous injection [23, 24], or to reduce side effects (e.g., itraconazole) [25] or to target specific organ in the body (e.g., the brain), [26-28] and local targeting e.g., ocular [29, 30] is well discussed in publications. However, very limited references are available on the application of nanosuspensions for dermal delivery [31, 32] to enhance penetration and efficiency of actives [18].

In one of the interesting study, *in vivo* and *in vitro* percutaneous absorption of apigenin was studied in three different vehicles like dimethyl sulphoxide (DMSO) and acetone/DMSO (4:1) and propylene glycol/DMSO (4:1). *In vitro* percutaneous absorption studies was performed in whole mouse skin, without subcutaneous fat, mounted on Franz diffusion cells at 37 °C showed penetration in order of (acetone/DMSO) > DMSO > (propylene glycol/DMSO). *In vivo* percutaneous absorption of apigenin on SENCAR mice showed saturation kinetics for DMSO or acetone /DMSO system and non-saturated absorption for propylene glycol/DMSO system. This information is very helpful basis while formulating dermal drug delivery system of apigenin [33].

Al Shaal et al., prepared apigenin nanosuspensions by smartCrystal technology (milling followed by homogenization) and scaled up to industrial level [34]. The size of apigenin crystal was reduced from 45 µm to 413 nm (D99% - 0.515 µm) with a polydispersity index of 0.202. The effect of preservatives on the stability of apigenin nanosuspensions was studied by author. The preservatives selected were 25% w/w ethanol, surfactant group (Plantacare® 2000, an alkyl polyglyceride at 5% w/w, water soluble vitamin E (TPGS, 2.5%) and cetyl pyridium chloride (CPC, 1%). Other well established synthetic preservatives like 1,3 pentandiol (hydrolite-5), triclosan, euxyl 9010 and multiEx were included in the study. Among the tested preservatives, multi EX, glycerol, triclosan, rokonsol and capryl glycol failed to maintain the stability of apigenin nanosuspensions. The zeta potential data predicted that preserved nanosuspensions using candidates from class alcohols and surfactants (used as preservatives) preserved suspensions well without causing any instability at long storage conditions. The same group showed improvement in solubility upon processing as nanosuspension of apigenin. Due to increase in saturation solubility further two fold increase in antioxidant activity was observed when analysed by DPPH method [35, 36]. The apigenin nanosuspensions were stable at 4, room temperature and 40 degree storage [34] and can be mixed in liquid state with any carrier like gel and creams. Apigenin nanosuspension was further converted into lyophilized form to produce dry form for tablet or capsules.

In another study, apigenin nanocrystals of 400-800 nm were prepared by supercritical antisolvent process and evaluated *in vivo* after single dose. In addition, the pharmacokinetic study of apigenin nanocrystals, in comparison to coarse powder, was also performed in rats after a single oral dose. *In vivo* studies in rats showed enhanced peak plasma levels (3.6-fold higher) and AUC (3.4-fold higher) compared to coarse apigenin powder [37]. Recently, newly developed ARTcrystal®-technology was applied for nanocrystals using three flavonoids viz. rutin, hesperidine and apigenin in comparison with commonly used high pressure homogenization technique. Apigenin nanocrystals produced by this technique were of 262 nm with polydispersity index of 0.31 which was comparable with convention HPH technique [38].

SELF (MICRO) EMULSIFYING DRUG DELIVERY (SMEDDS)

SMEDDS are microemulsions of size 100 to 250 nm composed of mainly surfactants and oil. SMEDDS systems are mainly employed to improve solubility, bioavailability, and to avoid enzymatic degradation. SMEDDS are prepared by dissolving drug in selected composition of oil, surfactant and co solvent. Several FDA approved ingredients are used in formulation of the SMEDDS. To enhance solubility and dissolution, apigenin is formulated as self-micro emulsifying drug delivery system using three components viz. polyethoxylated castor oil (Cremophor(®)EL), Highly purified diethylene glycol monoethyl ether (Transcutol(®)HP) and Propylene glycol monocaprylate (type II) (Capryol™ 90). The solubility of apigenin in SMEDDS was increased to about 15 mg/g. Apigenin loaded SMEDDS were spherical in nature and showed mean particle size of 17 nm [39].

Solid Dispersion

Solid dispersion is one of the widely used techniques to improve solubility and ultimately bioavailability of active compound. In this technique, mainly uniform dispersion of drug in solid state is obtained into water-soluble carrier. The dispersion is produced by directly melting drug in to carriers like polyethylene glycol, surfactants. The production technique includes melt evaporation method, melt extrusion, electro spinning and melt agglomeration. Recently, other techniques like supercritical technology and co-precipitation are being employed for production of solid dispersion. Apigenin solid dispersion was prepared with carbon nanopowder. Drug release profiles showed enhanced dissolution property of apigenin from solid dispersion compared to coarse powder. *In vivo* experiments in rats showed 1.83 times high area under the curve for solid dispersion compared to coarse apigenin. Preliminary toxicity study showed the developed drug delivery system is safe to use and has no added toxicity [40].

Ethosomes

Ethosomes are composed of phospholipid, ethanol and water. Various hydrophobic synthetic and herbal actives were encapsuled in ethosomes. Ethosomes offer advantage of improve penetration through different skin layers when applied dermally. Apigenin ethosomes were produced for topical use and their skin deposition and transdermal flux was tested against liposomes by Shen et al. The amount of phospholipids and short-chain alcohols (propylene glycol and ethanol) used in composition had profound effect on efficiency of ethosomes. Ethosomes showed marked differences in both *in vitro* and *in vivo deposition studies when compared to liposomes and deformable liposomes.* Ethosomes are ideal carrier to use against UV radiation induced skin inflammations as they showed marked reduction in cyclooxygenase-2 levels when tested against ultraviolet B (UVB) light induced inflammation in mouse skin [41].

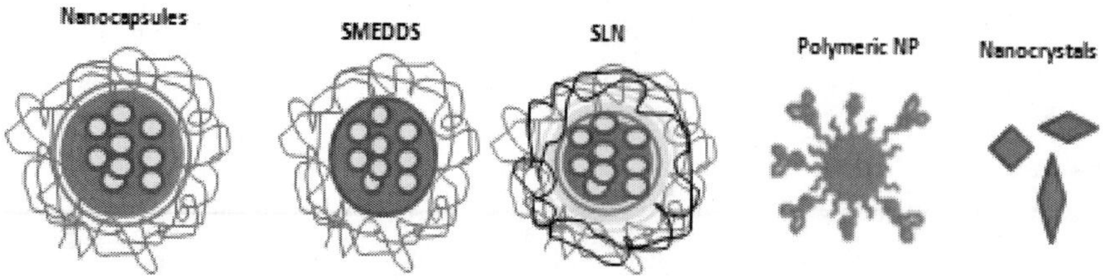

Figure 2. Different drug delivery platforms. Nanocapsules: drug encapsuled in oil core surrounded by polymeric or lipid layer. SMEDDS: drug dispersed in oil phase stabilized by surfactants, SLN: drug dispersed in lipid core (NLC-nanostructured lipid carrier- composed of solid lipid and liquid lipid core), Polymeric nanoparticles: drug encapsulated in polymeric core. NanoCrystals: surfactant stabilized nanonized drug particles.

Lipid Nanocapsules

Lipid nanocapsules are core-shell structures composed of an oil-filled core protected by polymeric or lipidic layer. They are mainly prepared by phase inversion method and holds promise in the delivery of hydrophobic drugs in the new therapeutic strategies. Nanocapsules are superior to nanoemulsions in terms of stability and release profile. The shell forms barrier between drug and biological environment thereby avoiding drug degradation. Among very limited study published on nanocapsules of apigenin, spherical lipid nanocapsules of apigenin were prepared by phase inversion method. The anti-proliferative activity of lipid nanocapsules inhibited growth rate of HepG2 cell and MCF-7 cell when tested by M assay [42].

Conclusion

Apigenin is promising candidate to medicine, food and cosmetics due to its variety of beneficial pharmacological effects. It uses diverse and multi molecular pathways to exert unique pharmacological actions *in vivo*. A very few strategies and nanotechnological delivery systems have been studied and reported in literature on improvement of the solubility and bioavailability of apigenin. The countable published studies based on *in vitro* data are success proof of developed drug delivery platforms. However, very limited data is available on *in vivo* performance of apigenin loded nanoparticle drug delivery systems. With no doubt nanotechnological drug delivery offers more functionality to apigenin by virtue of improved solubility.

References

[1] Lee, J.H., et al., Anti-inflammatory mechanisms of apigenin: Inhibition of cyclooxygenase-2 expression, adhesion of monocytes to human umbilical vein endothelial cells, and expression of cellular adhesion molecules. *Archives of Pharmacal Research,* 2007. 30(10): p. 1318-1327.

[2] Romanova, D., et al., Study of antioxidant effect of apigenin, luteolin and quercetin by DNA protective method. *Neoplasma,* 2001. 48(2): p. 104-7.
[3] Ghayur, M.N., H. Khan, and A.H. Gilani, Antispasmodic, bronchodilator and vasodilator activities of (+)-catechin, a naturally occurring flavonoid. *Arch Pharm Res,* 2007. 30(8): p. 970-5.
[4] Datta, B.K., et al., Analgesic, antiinflammatory and CNS depressant activities of sesquiterpenes and a flavonoid glycoside from Polygonum viscosum. *Pharmazie,* 2004. 59(3): p. 222-5.
[5] Czeczot, H., et al., Isolation and studies of the mutagenic activity in the Ames test of flavonoids naturally occurring in medical herbs. *Mutat Res,* 1990. 240(3): p. 209-16.
[6] Müller, R.H., B.H.L. Böhm, and G. M.J., Nanosuspensionen: Formulierungen für schwerlösliche Arzneistoffe mit geringer Bioverfügbarkeit. 2. Mitteilung: Stabilität, biopharmazeutische Aspekte, mögliche Arzneiformen und Zulassungsfragen. *Pharm. Ind.,* 1999. 61(2): p. 175-178.
[7] Rabinow, B.E., Nanosuspensions in drug delivery. *Nat Rev Drug Discov,* 2004. 3(9): p. 785-796.
[8] Stegemann, S., et al., When poor solubility becomes an issue: From early stage to proof of concept. *European Journal of Pharmaceutical Sciences,* 2007. 31(5): p. 249-261.
[9] Min et al., Measurement and correlation of solubilities of apigenin and apigenin 7-O-rhamnosylglucoside in seven solvents at different temperatures. *J. Chem. Thermodyn,* 2011. 43(3): p. 240-243.
[10] Sawant S and Shegokar R., Cancer research and therapy: Where are we today? *Int J Cancer Ther Oncol* 2014. 2(4): p. 02048.
[11] Das, S., et al., Strategic formulation of apigenin-loaded PLGA nanoparticles for intracellular trafficking, DNA targeting and improved therapeutic effects in skin melanoma in vitro. *Toxicology Letters,* 2013. 223(2): p. 124-38.
[12] Das, S., et al., Efficacy of PLGA-loaded apigenin nanoparticles in Benzo[a]pyrene and ultraviolet-B induced skin cancer of mice: mitochondria mediated apoptotic signalling cascades. Food and chemical toxicology : *An international journal published for the British Industrial Biological Research Association,* 2013. 62: p. 670-80.
[13] Zhai, Y., et al., Preparation and in vitro evaluation of apigenin-loaded polymeric micelles. Colloids and Surfaces A-Physicochemical and Engineering Aspects, 2013. 429: p. 24-30.
[14] Munyendo, W.L., et al., Micelles of TPGS modified apigenin phospholipid complex for oral administration: preparation, in vitro and in vivo evaluation. *Journal of Biomedical Nanotechnology,* 2013. 9(12): p. 2034-47.
[15] Buckton, G. and A.E. Beezer, The Relationship between Particle-Size and Solubility. *International Journal of Pharmaceutics,* 1992. 82(3): p. R7-R10.
[16] Hecq, J., et al., Preparation and characterization of nanocrystals for solubility and dissolution rate enhancement of nifedipine. *International Journal of Pharmaceutics,* 2005. 299(1-2): p. 167-177.
[17] Jinno, J., et al., Effect of particle size reduction on dissolution and oral absorption of a poorly water-soluble drug, cilostazol, in beagle dogs. *Journal of Controlled Release,* 2006. 111(1-2): p. 56-64.

[18] Keck, C.M. and R.H. Müller, *smartCrystals-Review of the Second Generation of Drug Nanocrystals, in Nanoparticlulates as Drug Carriers,* V.P. Torchilin, Editor 2008, Imperial College Press: London.
[19] Müller, R.H., C. Jacobs, and O. Kayser, *Nanosuspensions for the Formulation of Poorly Soluble Drugs, in Pharmaceutical Emulsions and Suspensions,* F. Nielloud and G. Marti-Mestres, Editors. 2000, Marcel Dekker. p. 383-407.
[20] Lai, F., et al., Diclofenac nanosuspensions: influence of preparation procedure and crystal form on drug dissolution behaviour. *Int J Pharm,* 2009. 373(1-2): p. 124-32.
[21] Mauludin, R., R.H. Muller, and C.M. Keck, Development of an oral rutin nanocrystal formulation. *Int J Pharm,* 2009. 370(1-2): p. 202-9.
[22] Van Eerdenbrugh, B., G. Van den Mooter, and P. Augustijns, Top-down production of drug nanocrystals: nanosuspension stabilization, miniaturization and transformation into solid products. *Int J Pharm,* 2008. 364(1): p. 64-75.
[23] Ganta, S., et al., Formulation and pharmacokinetic evaluation of an asulacrine nanocrystalline suspension for intravenous delivery. *Int J Pharm,* 2009. 367(1-2): p. 179-86.
[24] Gao, L., et al., Studies on pharmacokinetics and tissue distribution of oridonin nanosuspensions. *Int J Pharm,* 2008. 355(1-2): p. 321-7.
[25] Rabinow, B., et al., Itraconazole IV nanosuspension enhances efficacy through altered pharmacokinetics in the rat. *International Journal of Pharmaceutics,* 2007. 339(1-2): p. 251-260.
[26] Kayser, O. and A.F. Kiderlen, Delivery strategies for antiparasitics. *Expert Opin Investig Drugs,* 2003. 12(2): p. 197-207.
[27] Keck, C.M., NanoCrystal® Technology: A formulation approach for poorly water soluble Compounds, in Particle Design for APIs and Drug Products2008: Brussels.
[28] Schöler, N., et al., Atovaquone nanosuspensions show excellent therapeutic effect in a new murine model of reactivated toxoplasmosis. *Antimicrobial Agents and Chemotherapy,* 2001. 45(6): p. 1771-9.
[29] Agnihotri, S.M. and P.R. Vavia, Diclofenac-loaded biopolymeric nanosuspensions for ophthalmic application. *Nanomedicine-Nanotechnology Biology and Medicine,* 2009. 5(1): p. 90-95.
[30] Kassem, M.A., et al., Nanosuspension as an ophthalmic delivery system for certain glucocorticoid drugs. *International Journal of Pharmaceutics,* 2007. 340(1-2): p. 126-133.
[31] Mishra, P.R., et al., Production and characterization of Hesperetin nanosuspensions for dermal delivery. *International Journal of Pharmaceutics,* 2009. 371(1-2): p. 182-9.
[32] Teeranachaideekul, V., et al., Development of ascorbyl palmitate nanocrystals applying the nanosuspension technology. *Int J Pharm,* 2008. 354(1-2): p. 227-34.
[33] Li, B. and D.F. Birt, In vivo and in vitro percutaneous absorption of cancer preventive flavonoid apigenin in different vehicles in mouse skin. *Pharmaceutical Research,* 1996. 13(11): p. 1710-5.
[34] Al Shaal, L., R.H. Müller, and R. Shegokar, smartCrystal combination technology--scale up from lab to pilot scale and long term stability. *Pharmazie,* 2010. 65(12): p. 877-84.

[35] Al Shaal, L., R. Shegokar, and R.H. Muller, Production and characterization of antioxidant apigenin nanocrystals as a novel UV skin protective formulation. *International Journal of Pharmaceutics,* 2011. 420(1): p. 133-140.
[36] Al Shaal, L., R. Shegokar, and R.H. Müller, Apigenin smartCrystals for novel UV skin protective formulations. *Int J Pharm* 2011. 420 (1): p. 133-140.
[37] Zhang, J., et al., Preparation of apigenin nanocrystals using supercritical antisolvent process for dissolution and bioavailability enhancement. European journal of pharmaceutical sciences : *Official journal of the European Federation for Pharmaceutical Sciences,* 2013. 48(4-5): p. 740-7.
[38] Scholz, P. and C.M. Keck, Flavonoid nanocrystals produced by ARTcrystal®-technology. *Int J Pharm,* 2014. doi:10.1016/j.ijpharm. 2014.11.008.
[39] Zhao, L., et al., Design and evaluation of a self-microemulsifying drug delivery system for apigenin. *Drug Development and Industrial Pharmacy,* 2013. 39(5): p. 662-9.
[40] Ding SM, et al., Enhanced bioavailability of apigenin via preparation of a carbon nanopowder solid dispersion. Int J Nanomedicine, 2014. 13(9).
[41] Shen, L.N., et al., Enhanced in vitro and in vivo skin deposition of apigenin delivered using ethosomes. *International Journal of Pharmaceutics,* 2014. 460(1-2): p. 280-8.
[42] Ding, B.Y., H. Chen, and C. Wang, Preparation and in vitro evaluation of apigenin loaded lipid nanocapsules. *J. Nanosci. Nanotechnol.,,* 2013. 13: p. 6546-6552.

In: Apigenin and Naringenin
Editor: Nilus M. Stacks
ISBN: 978-1-63463-987-3
© 2015 Nova Science Publishers, Inc.

Chapter 4

UPDATE ON PROPERTIES OF APIGENIN RELATED TO INFLAMMATORY PROCESS

Gorzalczany Susana*
Pharmacology Department,
School of Pharmacy and Biochemistry,
University of Buenos Aires, Argentina

ABSTRACT

Inflammation is a key player in numerous human diseases. Several evidences support the relationship between inflammation and cancer. This is based on the finding that tumors often arise at sites of chronic inflammation. Natural flavonoids have been used to treat various inflammatory diseases. In particular, apigenin is a flavonoid that is widely distributed in dietary plants, especially in parsley, thyme, peppermint, olives, and herbs like chamomile. This review summarizes the current knowledge on the effects of apigenin on inflammation process, with a focus on the cellular and molecular mechanism involved. In this sense, different molecular and cellular targets are analyzed. Few clinical trials have described the relevance of apigenin in human diseases; the need for further controlled intervention trial will be discussed. Finally, taking into account its poor aqueous solubility leading to a poor absorption in gastrointestinal tract, novel strategies to improve its solubility and bioavailability, such as polymeric micelles, ethosomes, liposomes, will be shown.

ABBREVIATONS

AP-1	activator protein-1
COX_2	cyclooxygenase 2
CXCR-4	C-X-C chemokine receptor type 4
DNA	deoxyribonucleic acid
HO-1	heme oxygenase-1

* E-mail: sgorza@ffyb.uba.ar.

IKK	IkappaB kinase
IP-10	interferon gamma-induced protein 10
iNOS	inducible isoform nitric oxide synthase
IL	interleukin
LPS	lipopolysaccharide
MCP-1	monocyte chemoattractant protein-1
MAPKs	mitogen activated protein kinases
MDC	myeloid dendritic cell
MMPs	matrix metalloproteinases
mRNA	messenger ribonucleic acid
NF-κB	nuclear factor-kappaB
PG	prostaglandin
ROS	reactive oxygen species
SMEDDS	self-microemulsifying drug delivery system
STAT	signal transducer and activator of transcription
TNFα	tumor necrosis factor-α
TPA	12-O-Tetradecanoylphorbol-13-acetate
TPGS	tocopherol polyethylene glycol succinate

INTRODUCTION

The relationship between inflammation and cancer was first postulated by Galenus around 1,800 years ago. He proposed that tumors could arise from inflammatory tissue injury and both tissues were pathophysiologically similar [1]. Centuries later in 1863, Virchow hypothesized that the origin of cancer was at sites of chronic inflammation, in part based on his hypothesis that some classes of irritants, together with the tissue injury and ensuing inflammation that they cause, enhance cell proliferation. Although it is now clear that proliferation of cells alone does not cause cancer, sustained cell proliferation in an environment rich in inflammatory cells, growth factors, activated stroma, and DNA-damage-promoting agents, certainly potentiates and/or promotes neoplastic risk [2]. In this sense, diverse agents trigger the inflammation associated with human cancer. Murine models have confirmed this hypothesis, especially in terms of gastric (*H. pylori* infection), liver (cholangitis) and colon (colitis) cancers. In these and other animal cancer models, the cells and mediators of chronic inflammation act as tumor promoters at distinct phases of malignant progression [3, 4, 5, 6, 7]. It is interesting that many intracellular signal transduction pathways activated by pro-inflammatory mediators lead to the activation of oncogenes, such as RAS, MYC and RET. On the other hand, inhibition of pro-inflammatory mediators, such as tumour necrosis factor alpha (TNF-α), interleukin-1 beta (IL-1β) also leads to a reduction in tumour development. Also, the inhibition of transcription factors associated with inflammation processes such as nuclear factor kappa-light-chain-enhancer of activated B cells (NF-κB) or signal transducer and activator of transcription 3 (STAT3) [8, 9] shows the similar effect.

Wattenberg (1975) was the first who demonstrated the principles of chemoprevention by plant constituents in animal models [10]. His studies are based on the large number of the emerging epidemiological evidence for protective effects of diets rich in fruits and vegetables against a variety of cancers. Indeed, increasing evidence suggests the importance of plant-based foods in the prevention of chronic disease, such as heart disease, cancer, stroke, diabetes, Alzheimer's disease, cataracts, and age-related function decline [11, 12, 13].

Clinically, chemoprevention is categorized as primary, secondary, or tertiary. Primary chemoprevention is suited for the general population and for those who may be at increased risk for disease. Secondary chemoprevention is intended for patients with premalignant lesions that may progress to an invasive disease. On the other hand, tertiary chemoprevention is targeted to prevent disease recurrence or additional primary disease in those individuals who have already endured potentially curative therapy [14]. A wide array of phytochemicals, has been attracted scientists due to their property in altering cell cycle control, angiogenesis and inflammation process, representing a treatment strategy with potentially lower toxic effects, for both primary and secondary chemoprevention, although with less evidence for tertiary chemoprotection.

Polyphenols are common components of plant foods; they comprise a wide variety of molecules that have a polyphenol structure, but also molecules with one phenol ring, such as phenolic acids. The main groups of polyphenols are: flavonoids, phenolic acids, stilbenes and lignans. Flavonoids share a common carbon skeleton of diphenyl propanes, two benzene rings joined by a linear three-carbon chain. They are divided into 6 subclasses, depending on the oxidation state of the central pyran ring: flavonols, flavones, flavanones, isoflavones, anthocyanidins and flavanols. More than 4000 flavonoids have been identified in plants, and the list is constantly growing, taking into account the occurrence of numerous substitution patterns in which primary substituents (as hydroxyl group) can be substituted (glycosylated), yielding in a greater variety of highly complex structures [15, 16].

Particularly, apigenin (5,7,4′-trihydroxyflavone) is a naturally occurring flavonoid belonging to the flavone subclass and abundantly presents in common fruits (oranges, apples, cherries, grapes), vegetables (onions, parsley, broccoli, sweet green pepper, celery, barley, tomatoes) and beverages (tea, wine) [17]. Apigenin was first structurally identified in 1900, and synthesized in 1939. In its natural form, apigenin is present in foods mostly as glycoside conjugates, which are more soluble in water than theirs pure form. These glycosides are efficiently hydrolyzed in vivo by bacterial enzymes in the human intestinal tract to the free flavonoid [18, 19, 20].

Apigenin and Inflammation

Inflammation is a complex process that involves extensive changes in cellular and molecular component of physiology. Although controlled inflammation is an important process required for an array of processes including tissue repair, wound healing and for defense against invading foreign pathogens; chronic and uncontrolled inflammation is harmful and has now been linked to a number of human ailments. Chronic inflammation acts as a key regulator of tumour promotion and progression by several mechanisms including accelerated cell proliferation, evasion from apoptosis, enhanced angiogenesis and metastasis. The mechanism for cancer development in the presence of chronic inflammation involves the continuous presence of cytokines, chemokines, reactive oxygen species (ROS), oncogenes, cyclo.oxygenase-2 (COX2), matrix metalloproteinases (MMPs), activation of NF-κB, and activator protein 1 (AP-1) [21].

The immune system plays an important role in the development and progression of inflammatory process. An analysis of the ever increasing number of reports directs attention to the special role of the cytokine network. The group of inflammatory cytokines is the most important group of compounds participating in the inflammatory process. Among the many representative of this group, the greatest importance is attributed to IL-1β, TNFα, IL-6, IL-15, and IL-17. At the same time, one should not underestimate the role of anti-inflammatory cytokines that may modulate an inflammatory response and act protectively on tissues. The main representatives of the group of anti-inflammatory cytokines are IL-4 and IL-10 [22].

Human tumor necrosis factor was originally described as a circulating factor that can cause necrosis of tumours but has since been identified as a key regulator of the inflammatory response. Nineteen different members of the TNF superfamily which interact with different receptors have been identified. Most members of this family exhibit pro-inflammatory activities, which play crucial role in the innate and adaptive immunity, cell proliferation, and apoptotic processes. These cytokines are produced by different kind of cells, including macrophages, monocytes, T-cells, smooth muscle cells, adipocytes, and fibroblasts [23, 24]. Kowalski (2005) showed that apigenin significantly decreased the number of TNFα –mRNA copies in lipopolysaccharide (LPS) activated J774.2 macrophages [25]. In concordance with this result, it was shown that also exhibited inhibitory effects on TNFα induced E-selectin expression in human endothelial cells, being selectins an important factor related to effective neutrophil recruitment [26]. However, apigenin was less potent in regulating TNFα expression in human THP-1 derived macrophages [27]. The production of this cytokine and IL-6 was inhibited in murine macrophage cells and mice induced by LPS. But, the inhibition in mice did not display the same pattern as cell culture, indicating the in vitro activity must be confirmed using appropriate animal models [28]. A potent inhibitory activity of IL-6 production in the TNF-α stimulated MG-63 cell [29], in Di-(3-ethylhexyl) phthalate stimulated human umbilical vein endothelial cell and also in in vivo test [30, 31] was described too. In the same sense, the flavone induced a significant inhibition of IL-1β in LPS activated J774.2 macrophages [32] and of IL-17 while produced an increase of anti-inflammatory cytokine IL-10 [33]. Related to IL-4, many studies revealed the capacity of apigenin to inhibit this cytokine, but it is important to know that IL-4 is critical for the synthesis of IgE in B cell, and its inhibition may therefore be effective in inhibiting allergy disease [34].

Several ILs induce the activation of transcription factors, including NF-κB and AP-1 that are involved in the control of a large number of normal cellular process, such as immune and inflammatory responses, cellular growth, and apoptosis. NF-kB is expressed in almost all cell types and tissues, and specific NF-kB binding sites are present in the promoters/enhancers of a large number of genes. The primary mechanism for regulating NF-kB is through inhibitory IkB proteins, and the kinase that phosphorylates IkBs, namely, the IkB kinase (IKK) complex, representing the canonical pathway induced by TNF, IL-1, or LPS. Phosphorylation of serine residues leads to IkB ubiquitination and subsequent proteosomal degradation, releasing NF-kB dimer, which can then translocate to the nucleus and induce transcription of target genes [35]. It was shown that inhibitory effect of apigenin on NF-kB transcriptional activity in macrophages is at least four times higher than the activity reported on epithelial cells. Also, apigenin reduce IKK activity induced by LPS, modulating phosphorylation of p65 subunit at Ser 536, without affecting NF-kB DNA binding activity [36, 37, 38].

Also, Activator protein one (AP1) is key downstream target of mitogen-activated protein kinase (MAPK) signaling. AP1 include jun and fos family members. They form jun-jun and jun-fos dimmers that interact with specific AP1 transcription factor consensus DNA binding elements in target genes to regulate its expression. In this sense, it was shown that apigenin inhibited matrix metalloproteinase-1 (MMP1) production in vitro by interfering with Ca^{2+} dependent MAPKs and AP1 signalling [39].

On the other hand, chemokines are also essential for the directional migration of leukocytes during normal and inflammatory processes. Interleukin-8 (IL-8) and monocyte chemoattractant protein-1 (MCP-1) are the best characterized members of the CXC and CC chemokine subfamilies, respectively. Chemokines are involved in chemotaxis of monocytes, lymphocytes, neutrophils, eosinophils, basophils, natural killer cells, dendritic cells, and endothelial cells [40]. It was shown that apigenin inhibited production of monocyte derived chemokine (MDC) and IP-10 (Interferon inducible protein-10) in human THP-1 monocyte cells stimulated by LPS and this effect was induced by blocking MAPK pathways. The inhibition of c-raf phosphorylation might be the target of apigenin's property [41]. Both MDC and IP-10 plays a pivotal role in recruiting Th2 and Th1 cells, respectively. Other chemokines, such as MCP-1, secreted by activated macrophages, was also modified by apigenin. The flavone diminished MCP-1 release in LPS activated J774.2 macrophages, reducing significantly the number of MCP-1 mRNA copies [42]. Moreover, apigenin suppressed CXCR4 expression in different cells (BAsT, SW480 cells), indicating not cell type specificity. Indeed, this effect did not occur at the post-translational level but transcriptional level [43].

Different enzymes play an important role in inflammatory process. In this sense, cyclooxygenases (COX) are key enzymes required for prostanglandin (PG) synthesis, converting the arachidonic acid (AA) released by phospholipase A2 into PGs. 12-O-tetradecanoylphorbol-13-acetate (TPA) induce release of AA in cell and this effect was inhibited by cell treatment with apigenin which reduced the COX2 expression by blocking signal transduction of Akt [44]. Similar effect was observed on inducible nitric oxide synthase (iNOS) and heme oxygenase-1 (HO-1) in inflammatory conditions [45, 46].

Furthermore, oxidative stress, consequence of an imbalance of prooxidants and antioxidants in the organism, has been rapidly gaining recognition as a key phenomenon in chronic diseases. Oxidative stress can activate a variety of transcription factors including NFkB, AP-1, which in turn could mediate most chronic diseases, suggesting that oxidative stress, chronic inflammation, and cancer are closely linked [47]. In this sense, apigenin attenuates 2-deoxy-D-ribose-induced cell damage in pancreatic cell via oxidative stress, reversing the deleterious effects on micochondrial membrane potential [48] and produce a genoprotective effect in H2O2-stressed RWPE-1 cells by reduction in reactive oxygen species levels [49]. However, prooxidative effects were attributed to apigenin in vitro. It was described, apoptosis mediating by ROS in rheumatoid fibroblast-like synociocytes, and prooxidant effect on HL-60 cell line [50, 51].

Thus, apigenin appears to have different biological effects in different cell types, therefore these controversial effects should be clarified in future researchs in order to know the real impact on human health.

FORMULATIONS

Oral bioavailability of apigenin is relatively low because of its low lipid (0.001–1.63 mg/mL in nonpolar solvents) and water (2.16 μg/mL in water) solubility, which has severely limited its clinical development. Therefore, it is necessary to develop new technologies or formulations to improve its bioavailability. According to the Biopharmaceutics Classification System, apigenin is a class II drug with high intestinal membrane permeability and poor solubility, which can be improved by increasing the dissolution rate of the drug [52]. Solid dispersion is widely used to enhance the dissolution rate of poorly water-soluble drugs. Solubility and dissolution rate of poorly water soluble drugs can be improved by reducing particle size, enhancing wettability and porosity, and changing the drug state from crystalline to amorphous. Carbon nanopowder was used to improve its bioavailability, drug release profiles showed that apigenin dissolution from the carbon nanopowder-aginenin system (weight ratio, 6:1) was improved by 275% compared with that of pure apigenin. Moreover, the pharmacokinetic analysis in rats showed that the apigenin area under the curve 0–t value was 1.83 times higher for the carbon nanopowder-apigenin system than for pure flavonoid, indicating that its bioavailability was significantly improved [53].

Also, nanocrystals are one of the safest and also most practical, industrially feasible ways to enhance the dissolution velocity, not only increases the dissolution velocity, but also increases the saturation solubility [54]. Nanocrystals can be produced using two different wet milling methods, low energy milling technology or high pressure homogenization. In low energy milling the grinding of the particles takes place by beads or pearls. High pressure homogenization forces the particles to pass through a tiny gap (5–10 m, depending on the applied pressure) with high velocity causing the particle size reduction by cavitation, particle collision and shear forces. A second generation of drug nanocrystals (smartCrystals) prepared by a combination of both technologies was introduced. This technology was used to obtain nanocrystals of apigenin, with long term stability at storage temperatures 4°C, room temperature and 40°C [55]. Also, Zhang et al. (2013) prepared nanocrystals by the supercritical anti solvent process [56]. After nanocrystals formation, it was observed no substantial change in crystalline structure but decreased particle size with smooth surface in spherical shape . The prepared apigenin nanocrystals exhibited much more rapid dissolution profiles than apigenin coarse powder in different dissolution media.

Ethosomes, a novel type of liposomes containing 20–50% of ethanol (v/v), can enhance drug loading by increasing apigenin's solubility and transdermal absorption. Ethosomes are lipid vesicles containing phospholipids, relatively high concentrations of lower alcohols (ethanol or propylene glycol), and water. These innovative vesicles are attractive because, owing to their high deformability, they can transport active agents through the stratum corneum into deeper layers of the skin more effectively than traditional liposomes and deformable liposomes [57].

In the same sense, mixed micelles were designed to increase oral bioavailability of flavonoid. The phospholipid complex technology was exploited alongside TPGS´stabilizing effect by inducing sterical hindrance of the phase II enzymes. This prevented extensive metabolism of apigenin while inhibiting P-glycoprotein´s exocytosis. TPGS modified micelles of apigenin-phospolipid complex were prepared by thin film hydration method. They were found to be stable for more than 90 days, and the micelles increased intestinal absorption 2.4 fold, comparing to free drug. Furthermore, this technology improved its pharmacological activity in S180 carcinoma mice [58]. Besides, polymeric micelles composed of Pluronic P123 and Solutol HS 15 were prepared by a thin-film dispersion method and demonstrated to be effective in inhibiting the growth of tumor cells [59]. So, this kind of formulation can become a potential nanocarrier of apigenin.

In adittion, self-microemulsifying drug delivery system (SMEDDS) was prepared to improve the solubility and dissolution of apigenin. The optimal formulation of SMEDDS obtained was comprised of 60% Cremophor® El, 30% Transcutol®HP and 10% Capryol™ 90. The equilibrium solubility was about 15 mg/g, and it could increase the solubility of apigenin in water for about 7500 folds [60].

Finally, poly (lactic-co-glycolide) is biodegradable, biocompatible and non-toxic polymers, and has satisfactory nano-encapsulation potentials, with smaller size and uniform spatial planar frequency. These properties give it the ability to enter cells and generate faster action; apigenin loaded in poly (lactic-co-glycolide) nanoparticle was tested. Das et al. (2013) showed that this formulation could be an effective drug delivery system, especially for cancer therapy [61].

Clinical Trials

Epidemiological studies considerably support the notion that diets rich in plant polyphenols are associated with a number of health benefits. Hertog et al. (1993) describe that average intake of flavonoids combined was 23 mg/day in adults in Netherlands [62]. The most important flavonoid was quercetin (mean intake 16 mg/day), followed by kaempferol. myricetin, luteolin, and apigenin. The most important sources of flavonoids were tea (48% of total intake), onions (29%), and apples (7%). Studies examining the chemopreventive activity of phytochemicals have shown widely varying results [63]. Wang et al. (2009) found no significant association between baseline intake of 5 common flavonoids (quercetin, kaempferol, myricetin, luteolin and apigenin) or some flavonoid rich foods and the risk of cancer in middle-aged and older, predominantly white women [64]. Although, other authors provided limited support for an association between intake of those 5 flavonoids and risk of ovarian cancer, these authors could demonstrate that only intake of apigenin could be associated with a significant decrease in ovarian cancer risk [65]. In the same sense, in a large study with 38,018 women, the high intake of these flavonols and flavones showed a reduction of development of type 2 diabetes, when biomarkers of insulin resistance and systemic inflammation was analyzed [66]. These differences show us that more precise assessments related to flavonoid intake´s benefit, using relevant biomarkers are necessary to confirm or refute their findings.

Although epidemiologic studies are useful for evaluation of the human health effects of long-term exposure to physiologic concentrations of polyphenols, it is necessary controlled randomized clinical trials to answer the question if data from animal models and *in vitro* systems are relevant for human disease outcomes. An important phenomenon to take into account is that, their bioactivity may be greatly influenced by absorption and for their extensive metabolism before reaching target organs.

Apigenin has shown anti-inflammatory and anti-oxidative properties, and it is strong candidate for a new treatment option for patients with diseases related to inflammatory process such as inflammatory bowel disease, rheumatoid arthritis and even cancer [67, 68, 69, 70].

In spite of great amount of preclinical studies using in vitro and in vivo methods, hardly any randomized clinical trials have been conducted. Phase 2 interventional´s study is in progress in order to obtain information about the potential prevention of neoplasia recurrence induced by dietary bioflavonoid supplementation (20 mg apigenin and 20 mg epigallocathecin) [71], but not related to inflammatory process.

Conflicting epidemiological inferences and discrepancies between *in vitro* and *in vivo* studies may be due to its erratic bioavailability. Aspects concerning these facts, but also related to dose levels, administration frequency and potential side effects remain to be addressed in future clinical trials.

TOXICITY

It is important that apigenin has very low intrinsic toxicity and therefore there has been no recommendation by the National Toxicology Program or Interagency Committee for Chemical Evaluation and Coordination to conduct toxicological studies and it is not listed in the Toxic Substance Control Act Inventory of the Environmental Protection Agency [72]. However, Singh et al. (2012) showed a significantly increased serum ALT; AST, ALP and LPO at doses of 100 and 200 mg/kg by intraperitoneal route in rats, without any effect at lower doses. This information could indicate a pro-oxidant activity and potential damage in the liver induced by apigenin [73]. In contrast to it, Ali et al. (2014) demonstrated the hepato-protective and anti-genotoxic effect of oral administration of apigenin in preclinical model of hepatotoxicity induced by N-nitrosodiethylamine in rats [74]. Therefore, further studies are required to confirm these results and to evaluate the impact on human health of it.

CONCLUSION

Numerous studies have been published identifying molecular mechanisms of apigenin effects. Mostly evidence of efficacy has been derived from animal models and in vitro test. The available information let us know the potential utility of apigenin for the possible treatment of diseases related to inflammatory process. Nevertheless, its use requires extensive clinical testing in order to optimize its potential use in patients. Another challenge will be to develop specific preparations in order to deliver apigenin to the target structures, enhancing its bioavalilabitlity. The research of compounds isolated from natural source, is now in a dynamic process and they could be a novel way to improve the currently treatment for diseases relate to inflammation process.

REFERENCES

[1] F. Okada, *Yonago Acta. Med.,* 57, 65 (2014).

[2] L. Coussens and Z. Werb, *Nature,* 420, 860 (2002).

[3] F. Greten, L. Eckman, T. F. Greten, J. M. Park, Z. Li, L. J: Egan and M. F. Kagnoff, *Cell,* 118, 285 (2004).

[4] E. Pikarsky, R. M. Porat, I. Stein, R. Abramovitch, S. Amit, S. Kasem, E. Gutkovich-Pyest, S. Uriell-Shoval, E. Galun, and Y. Ben-Neriah, *Nature,* 431, 4461 (2004).

[5] F. Balkwill, C. Kellie, A. Mantovani, *Cancer Cell,* 7, 211 (2005).

[6] Y. Wu, S. Antony, J. L. Meitzler, J. H. Doroshow, *Cancer Lett.,* 345, 164 (2014).

[7] L. Qiao, X. Li, *Cancer Lett.,* 345, 150 (2014).

[8] B. Müller - Hübenthal, M. Azemar, D., Lorenzen, M. Huber, M. Freudenberg, C. Galanos, C. Unger and B. Hildenbrandsky, *Anticancer Res.*, 29, 4795 (2009).

[9] C. Van Waes, *Clin. Cancer Res.,* 13, 1076 (2007).

[10] L. W. Wattenberg L. W., *Cancer Res.,* 35, 3326 (1975).

[11] G. Block, B. Patterson and A. *Nutrition and Cancer,* 18, 1 (1992).

[12] I. Johnson, *Proceedings of the Nutrition Society*, 66, 207 (2007).

[13] R. Liu, *Adv. Nutr.,* 4, 384S (2013).

[14] K. Landis-Piwowar and N. R. Iyer, *Cancer Growth and Metastasis*, 7, 19 (2014).

[15] M. D'Archivio, C. Filesi, R. Di Benedetto, R. Gargiulo, C. Giovannini and R. Masella, *Ann. Ist Super Sanità.*, 43, 348 (2007).

[16] R. Tsao, *Nutrients,* 2, 1231 (2010).

[17] D. Patel, S. Shukla, S. Gupta, *Int. J. Oncol.*, 30, 233 (2007).

[18] J. A. Ross, C. M. Kasum, *Annu. Rev. Nutr.*, 22, 19 (2002).

[19] B., Li, D. H. Robinson, D. F. Birt, *J. Pharm. Sci.*, 86, 721 (1997).

[20] X, Tong and J. Pelling, *Anticancer Agents Med. Chem.*, 13, 971 (2013).

[21] G. Sethi, M. Shanmugam, L. Ramachandran, A. Kumar and V. Tergaonkar, *Biosci. Rep.,* 32, 1 (2012).

[22] P. Wojdasiewicz, A. Poniatowski and D. Szukiewicz, *Mediat. Inflamm.,* Article ID 561459 http://dx.doi.org/10.1155/2014/561459 (2014).

[23] C. Popa, M. G. Netea, P. van Riel, J. van der Meer, and A. F. Stalenhoef, *J. Lip. Res.,* 48, 751 (2007).

[24] S. C. Gupta, A. K. Tyagi, P. Deshmukh-Taskar, M. Hinojosa, S. Prasad, B. B. Aggarwal, *Arch. Biochem. Biophys.*, 559, 91 (2014).

[25] J. Kowalsi, A. Samojedny, M. Paul, G. Pietsz, T. Wilczok, *Pharmacol. Rep.*, 57, 390 (2005).

[26] Y. Takano-Ishikawa, M., Goto, K. Yamaki, *Phytother. Res.,* 17, 1224 (2003).

[27] X. Zhang, G. Wang, E., Gurley and H. Zhou, *Plos one*, 9, e107072 (2014).

[28] A. T. Smolinski, J. J. Pestka, *Food Chem. Toxicol.*, 41, 1381 (2003).

[29] A. R. Kim, Q., Jin, H. G. Jin, H. J. Ko, E. R. Woo, *Arch. Pharm. Res.,* 37, 845 (2014).

[30] J. Wang, Y. Liao, J. Fan, T. Ye, X. Sun, S. Dong, *Inflammation,* 35, 1466 (2012).

[31] E. M. Drummond, N. Harbourne, E. Marete, D. Martyn, J. Jacquier, D. O'Riordan, E. Gibney, *Phytother. Res.,* 27, 588 (2013).

[32] J. Kowalski, A. Samojedny, M. Paul, G. Pietsz, T. Wilczok, *Pharmacol. Rep.*, 57, 390 (2005).

[33] H. Urushima, J. Nishimura, T. Mizushima, N. Hayashi, K. Maeda, T. Ito, *Am. J. Physiol. Gastrointest. Liver Physiol.*, Oct 30. doi: 10.115/ ajpgi.00294.2014 (2014).
[34] S. Yamashita, T. Yamashita, K. Yamada, H. Tachibana, *Febs Lett.*, 584, 775 (2010).
[35] A. Oeckinghaus and S. Ghosh, *Cold Spring Harb. Perspect. Biol.*, 1:a000034 (2009).
[36] B. Nicholas, S. Batra, M. Vargo, O. Voss, M. Gavrilin, M. Wewers, D. Guttidge, E. Grotewold and A. Doseff, *J. Immunol.*, 179, 7121 (2007).
[37] O. H. Kang, J. H. Lee, D. Y. Kwon, *Immunopharmacol. Immunotoxicol.*, 33, 473 (2011).
[38] D. G. Wu, P. Yu, J. W. Li, P. Jiang, J. Sun, H. Z. Wang, L. D. Zhang, M. B. Wen, P. Bie, *Toxicol. Lett.*, 224, 157 (2014).
[39] Y. Hwang, K. Oh, H. Yun, H. Jeong, *J. Dermatol. Sci.*, 61, 23 (2011).
[40] P. Proost, A. Wuyts, J. van Damme, *J. Clin. Lab. Res.*, 26, 211 (1996).
[41] C. H. Huang, P. L. Kuo, Y. L., Hsu, T. T. Chang, H. I. Tseng, Y. T. Chu, C. H. Kuo, H. N. Chen, C. H. Hung, *J. Med. Food*, 13, 391 (2010).
[42] J. Kowalski, A. Samojedny, M. Paul, G. Pietsz, *Wiad. Lek.*, 59, 634 (2006).
[43] L. Wang, L. Kuang, J. A. Hitron, Y. O. Son, X. Wang, A. Budhraja, J. Lee, P. Poyil, G. Chen, Z. Zhang, J. Luo, X. Shi, *Toxicol. Appl. Pharmacol.*, 272, 1 (2013).
[44] R. T. Van Dross, X. Hong, J. C. Pelling, *Mol. Carcinog.*, 44, 83 (2005).
[45] G. Jeong, S. Lee, S. Jeong, Y. Kim, E. Kim, *Int. Immunopharmacol.*, 9, 1374 (2009).
[46] A. Abate, G. Yang, R. Wong, H. Schroder, D. Stevenson, P. Dennery, *Free Radic. Biol. Med.*, 39, 711 (2005).
[47] S. Reuter, S. Gupta, M. Chaturvedi, B. Aggarwal, *Free Radic. Biol. Med.*, 49, 1603 (2010).
[48] K. Suh, S. Oh, J. Woo, S. Kim, J. Kim, Y. Kim, S., Chon, *Biol. Pharm. Bull.*, 35, 121 (2012).
[49] H. Shama, R. Kanwal, N. Bhaskaran, S. Gupta, *Plos One*, 9, e 91588 (2014).
[50] G. Shin, C. Kim, J. Lee, W. Cho, S. Lee, M. Jeong, J. Cho, K. Lee, *Chem. Biol. Interact.*, 182, 29 (2009).
[51] N. Miyoshi, K. Namura, T. Yamada, T. Osawa, Y. Nakamura, *Arch. Biochem. Biophys.*, 466, 274 (2007).
[52] S. Stegemann, F. Leveiller, D. Franchi, H. de Jong, H. Linden, *Eur. J. Pharm. Sci.*, 31, 249 (2007).
[53] S. Ding, Z. Zhang, J. Song, X. Cheng, J. Jiang, X. Jia, *Int. J. Nanomed.*, 9, 2327 (2014).
[54] C. M. Keck and R. H. Müller, In: V. P. Torchilin (ed.) Nanoparticlulates as Drug Carriers. London, Imperial College Press (2008).
[55] L. Al Shaal, R. H. Müller, R. Shegokar, *Pharmazie*, 65, 877 (2010).
[56] J. Zhang, Y. Huang, D. Liu, Y. Gao, S. Qian, *Eur. J. Pharm. Sci.*, 48, 740 (2013).
[57] L. Shen, Y. Zhang, Q. Wang, L. Xu, N. Feng, *Int. J. Pharm.*, 460, 260 (2014).
[58] W. Munyendo, Z. Zhang, S. Abbad, A. Waddad, H. Lu, L. Baraza, J. Zhou, *J. Biomed. Nanotechnol.*, 9, 2034 (2013).
[59] Y. Zhai, S. Guo, C. Liu, C. Yang, J. Dou, L. Li, G. Zhai, *Colloids Surfaces A.*, 429, 24 (2013).
[60] L. Zhao, L. Zhang, L. Meng, J. Wang, G. Zhai, *Drug Dev. Ind. Pharm.*, 39, 662 (2013).
[61] S. Das, J. Das, A. Samadder, A. Paul, A. Khuda-Bukhsh, *Toxicol. Lett.*, 223, 124 (2013).
[62] M. Hertog, P. Hollman, M. Katan, D. Kromhout, *Nutr. Cancer*, 20, 21 (1993).

[63] M. Uzarska, R. Czajkowski, R. Schwartz, A. Bajek, B. Zegarska, T. Drewa, *Melanoma Res.,* 23, 426 (2013).
[64] L. Wang, I. Lee, S. Zhang, J. Blumberg, J. Buring, H. Sesso, *Am. J. Clin. Nutr.*, 89, 905 (2009).
[65] M. Gates, A. Vitonis, S. Tworoger, B. Rosner, L. Titus-Ernstoff, S. Hankinson, D. Cramer, *Int. J. Cancer*, 124, 1918 (2009).
[66] Y. Song, J. Manson, J. Buring, H. Sesso, S. Liu, *J. Am. Coll. Nutr.,* 24, 376 (2005).
[67] A. Ganjare, S. Nirmal, A. Patil, *Fitoterapia,* 82, 1052 (2011).
[68] G. Shin, C. Kim, J. Lee, W. Cho, S. Lee, M. Jeong, J. Cho, K. Lee, *Chem. Biol. Interact.*, 182, 29 (2009).
[69] D. Ravishankar, A. Rajora, F. Greco, H. Osborn, *Int. J. Biochem. CB.,* 45, 2821 (2013).
[70] H. Hoensch, B. Groh, L. Edler, W. Kirch, *World J. Gastroenterol.*, 14, 2187 (2008).
[71] https://www.clinicaltrials.gov/ct2/show/record/NCT00609310?term=apigenin&rank=1
[72] K. Rithidech, M. Tungjai, P. Reungpatthanaphong, L. Honikel, S. Simon, *Mutat. Res.,* 749, 29 (2012).
[73] P. Singh, S. Mishra, S. Noel, S. Sharma, S. Rath, *Plos One*, 7, e31964 (2012).
[74] F. Ali, Rahul, F., Naz, S. Jyoti, Y. Siddique, *Mutat. Res. Genet. Toxicol. Environ. Mutagen.*, 767, 13 (2014).

In: Apigenin and Naringenin
Editor: Nilus M. Stacks

ISBN: 978-1-63463-987-3
© 2015 Nova Science Publishers, Inc.

Chapter 5

INHIBITORY EFFECTS OF APIGENIN ON MAMMALIAN DNA POLYMERASE AND PROLIFERATION OF HUMAN CANCER CELLS

***Yoshiyuki Mizushina*[1,2*], *Kazuaki Shiomi*[1], *Takefumi Onodera*[1,2] *and Isoko Kuriyama*[1]**

[1]Laboratory of Food and Nutritional Sciences, Faculty of Nutrition, Kobe Gakuin University, Nishi-ku, Kobe, Hyogo, Japan

[2]Cooperative Research Center of Life Sciences, Kobe Gakuin University, Chuo-ku, Kobe, Hyogo, Japan

ABSTRACT

In this review, inhibitory activity against mammalian DNA polymerases (pols) by ten major plant flavonoids was investigated. Apigenin (4′,5,7-trihydroxyflavone) was the most potent inhibitor of pols among the compounds tested. Apigenin did not affect the activity of plant (cauliflower) pol α or prokaryotic pols, such as the Klenow fragment of pol I from *Escherichia coli*, *Taq* pol and T4 pol. Apigenin did not inhibit the activities of other DNA metabolic enzymes, such as mouse IMP dehydrogenase (type II), T4 polynucleotide kinase and bovine deoxyribonuclease I. Apigenin also did not influence direct binding to double-stranded DNA as determined by thermal transition analyses. The ten major flavonoids were investigated as to whether they could prevent proliferation of human colon HCT116 carcinoma cells. Suppression of growth of human cancer cells showed the same tendency as inhibition of activity of mammalian pols. Apigenin showed the strongest suppression of proliferation of human cancer cells. These results suggest that the decrease of proliferation of cancer cells by flavonoids containing apigenin could be based on the inhibition of cellular pols. Apigenin did not suppress the growth of human normal cells (human dermal fibroblasts), and induced apoptosis in human cancer (HCT116) cells. Therefore, apigenin could be used as an anti-cancer agent, and apigenin-enriched fractions could be used in functional foods for cancer prevention.

* Corresponding author: Yoshiyuki Mizushina, PhD. E-mail: mizushin@nutr.kobegakuin.ac.jp; Tel.: +81-78-974-1551 (ext.3232); Fax: +81-78-974-5689.

Keywords: Flavonoids, apigenin, DNA polymerase, apoptosis, anti-cancer

ABBREVIATIONS

BSA	bovine serum albumin
DMSO	dimethyl sulfoxide
dNTP	2′-deoxyribonucleoside 5′-triphosphate
dsDNA	double-stranded DNA
dTTP	2′-deoxythymidine 5′-triphosphate
HDF	human dermal fibroblasts
IC_{50}	Half-maximal inhibitory concentration
IMP	inosine-5'-monophosphate
K_i	inhibition constant
K_m	Michaelis constant
LD_{50}	Fifty-percent lethal dose
Pol	DNA-dependent DNA polymerase (E.C. 2.7.7.7)
TdT	terminal deoxynucleotidyl transferase
Tm	melting temperature
TUNEL	terminal deoxynucleotidyl transferase dUTP nick end labeling
V_{max}	maximum velocity

1. INTRODUCTION

Cancer is a major public-health burden in Japan and other "developed" countries, resulting in ≈7 million deaths worldwide every year [1]. More than 11 million people are diagnosed with cancer every year, and it is estimated that there will be 16 million new cases per year by 2020 [1]. The most common cancers on a global scale for men and women are cancers of the lung, colon/rectum and stomach. Among men, cancer of the lung, colon/rectum and prostate gland are the most common cancers worldwide. For women, the most common cancers besides those of the lung and colon/rectum are those of the breast and cervix [2, 3].

Developments in cancer treatment have led to significantly improved survival and quality-of-life for cancer patients in the past three decades. Presently, the major treatment modalities for cancer are chemotherapy, surgery and radiation therapy [4–6]. In recent years, immunotherapy has also been pursued in cancer management [7]. Chemotherapy is one of the most effective treatments available for cancer [4–6]. New drugs and therapeutic strategies are being researched and developed continuously but the status of chemotherapy is far from satisfactory [8]. With some notable exceptions, the efficacy of chemotherapy is limited and severe drug-related side effects are common. It is believed that prolonged chemotherapy treatment weakens the immunologic defense system of the body, and leaves patients susceptible to other diseases and infections. Surgery is the least harmful conventional form of cancer treatment but not all cancers can be resected [9]. Radiation therapy is another treatment option for cancer, but has several potentially harmful side-effects, including weakened resistance to other diseases and the potential to be carcinogenic in itself [10]. Therefore, there is an urgent need to develop mechanism-based approaches for cancer management. The practical goals of this approach should be to decrease the incidence of invasive cancer and deaths from cancer at an early age through pharmacologic interventions reliant upon prevention rather than cure. Such an intervention is termed "chemoprevention" [11–14].

Chemoprevention is a rapidly growing area of oncology that focuses on the prevention of cancer using naturally occurring or synthetic agents [11–14]. This pharmacologic approach relies on the identification of healthy individuals deemed to be at a higher risk of developing cancer, and for whom a pharmacologic agent can inhibit the onset of cancer effectively. In addition to inhibiting or delaying the onset of neoplasia by blocking neoplastic inception, chemoprevention plays a part in the prevention of the development of invasive and metastatic properties in established neoplasms [11–15]. Chemoprevention of cancer, therefore, differs from cancer treatment in that the goal of this approach is to lower the prevalence of cancer. The cancer-inhibitory effects of various nutrients derived from plants as well as of non-nutritive plant-derived constituents ("phytochemicals") have been confirmed in various cell-culture systems and animal models of cancer [15, 16]. Such evidence has led to an increased emphasis on cancer-prevention strategies in which these dietary factors are utilized. Two major diet-related cancer prevention strategies have evolved, cancer chemoprevention and dietary cancer prevention, with appreciable overlap existing between them. In general, cancer chemoprevention involves pharmacologic intervention with synthetic or naturally occurring compounds to prevent, inhibit or reverse carcinogenesis or stop the development of invasive cancer [11–14]. In contrast, diets containing an abundance of fruits and vegetables and other plant-derived agents protect against various diseases, including epithelial cancers [17, 18]. Epidemiologic studies, including several case–control and cohort studies, have provided data that overwhelmingly support an inverse association between intake of fruits and vegetables and cancer risk [19–21]. These nutritional practices are effective types of dietary cancer prevention. Numerous components found in fruit and vegetables may contribute to their effectiveness in reducing the risk of cancer, including micronutrients, dietary fiber and various polyphenolic agents [15–17].

The principal plant-derived agents thought to provide protection against cancer are flavonoids and dietary fiber [22, 23]. Flavonoids are the most common and widely distributed polyphenolic compounds, present ubiquitously in foods of plant origin [24, 25]. Flavonoids comprise ≈5,000 compounds that are defined chemically as substances composed of a common phenylchromanone structure (C6-C3-C6), with one or more hydroxyl substitutes [24, 25]. These are, in general, classified into flavanones, flavones and isoflavones (Figure 1).

Flavonoids have been demonstrated to exert various biologic effects in numerous mammalian systems *in vitro* and *in vivo*, acting as free radical-scavengers and anti-oxidants, as well as exhibiting anti-mutagenic, anti-inflammatory, anti-viral, and purgative effects [25]. Additionally, other biologic effects induced by flavonoids include lowering the plasma levels of low-density lipoproteins, inhibiting platelet aggregation, and reducing cell proliferation [23, 25–27]. These effects are related to their actions in inhibiting the cell cycle, diminishing oxidative stress, improving the efficacy of detoxification enzymes, inducing apoptosis, and stimulating the immune system [25–27]. These inherent properties of flavonoids categorize them as a class of beneficial compounds possessing health-promoting and disease-preventing dietary effects, including efficacy in cancer prevention. An added benefit is that these compounds are associated with very low toxicity, making them excellent choices for chemoprevention protocols.

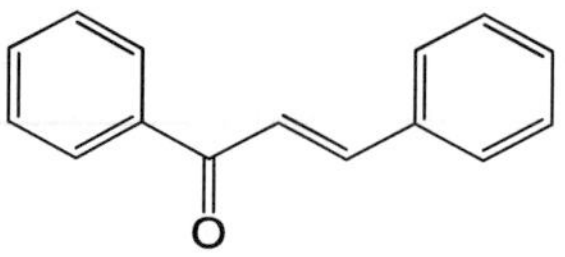

Chalcone (**1**)

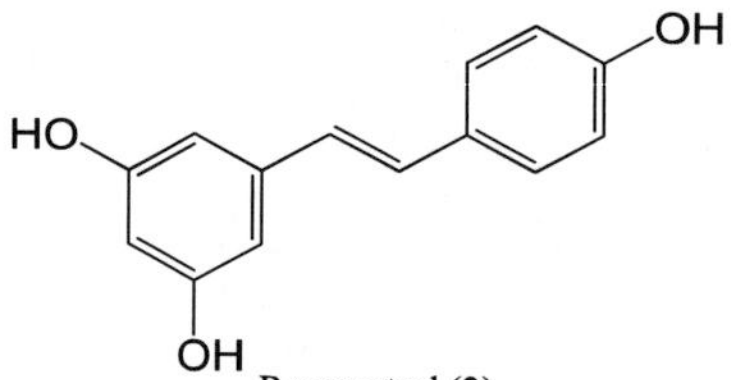

Resveratrol (**2**)

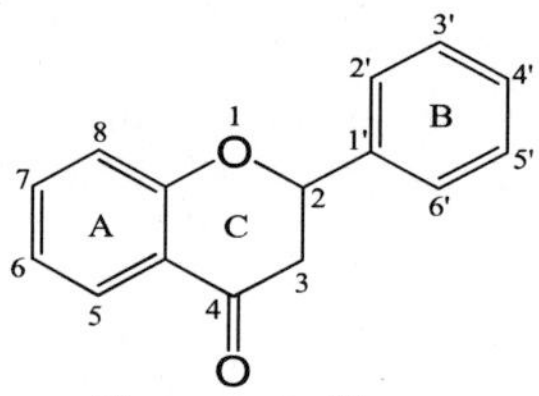

Flavanone backbone

	3	5	7	3'	4'	5'
Flavanone (**3**)	H	H	H	H	H	H
Naringenin (**4**)	H	OH	OH	H	OH	H
Hesperetin (**5**)	H	OH	OH	OH	OCH_3	H

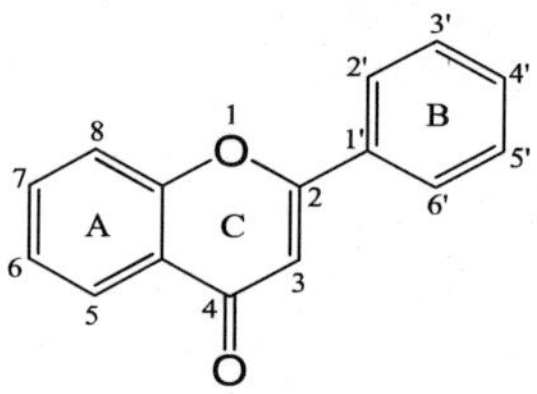

Flavone backbone

	3	5	7	3'	4'	5'
Flavone (**6**)	H	H	H	H	H	H
Apigenin (**7**)	H	OH	OH	H	OH	H
Apigenin-7-*O*-glucoside (**8**)	H	OH	Glu	OH	OH	H

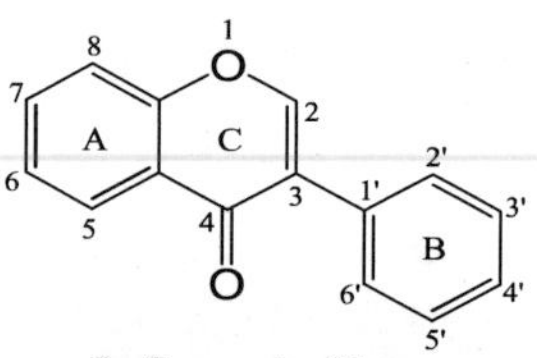

Isoflavone backbone

	3	5	7	3'	4'	5'
Daidzein (**9**)	—	H	OH	H	OH	H
Genistein (**10**)	—	OH	OH	H	OH	H

Figure 1. Structure of ten plant flavonoids. Chalcone (**1**), resveratrol (**2**), flavanones (**3–5**), flavones (**6–8**) and isoflavones (**9** and **10**) are shown. Glu: glucose.

2. EUKARYOTIC DNA POLYMERASE (POL) SPECIES AND SCREENING OF THEIR INHIBITORS

Pol (DNA-dependent DNA polymerase, E.C. 2.7.7.7) catalyzes the polymerization of deoxyribonucleotides along a DNA strand that is "read" and used as a template [28]. The newly polymerized molecule is complementary to the template strand and identical to the non-template strand. Pol can add free nucleotides only to the 3′ end of a newly formed strand, resulting in elongation of the new strand in the 5′→3′ direction.

The human genome encodes ≥15 pols which function in cellular DNA synthesis [29–32]. Eukaryotic cells contain three replicative pols (α, δ, ε), one mitochondrial pol (γ), and ≥11 non-replicative pols (β, ζ, η, θ, ι, κ, λ, μ, ν, terminal deoxynucleotidyl transferase (TdT), REV1) [29–32]. Pols have a highly conserved structure, with their overall catalytic subunits showing little variance among species; conserved enzyme structures are usually preserved over time because they undertake important cellular functions that confer evolutionary advantages.

On the basis of sequence homology, eukaryotic pols can be divided into four main families: A, B, X and Y [29–32]. Family A includes mitochondrial pol γ, as well as pols θ and ν; family B includes the three replicative pols α, δ and ε, and also pol ζ; family X comprises pols β, λ and μ as well as TdT; and family Y includes pols η, ι and κ, in addition to REV1.

Pols have important roles in the maintenance of key eukaryotic systems, such as the replication, recombination and repair of DNA [33]. Hence, pol inhibitors can be employed as anti-cancer chemotherapy agents to inhibit cell proliferation. Based on the strategic effects of pol inhibitors, we established an assay for determining pol inhibitors [34]. We have screened mammalian pol inhibitors in natural products, including food materials and chemically synthesized materials, for 20 years. More than 100 compounds have been identified as pol inhibitors [35, 36], and we have identified some flavonoids from the Japanese vegetable *fuki-noto* (edible flower bud of the fuki (*Petasites japonicas*) plant) [37] and green tea (*Camellia sinensis*) leaves [38], which have inhibitory activity against mammalian pol species.

In this review, we focused on the inhibitory effect of flavonoids (prepared compounds **1–10** in Figure 1) on mammalian pols. We also investigated the chemical constituents responsible for inhibiting pol activities and their cancer-preventive effects.

3. Effects of Flavonoids on Pol Activity

3.1. Preparation of Pols

Pol α was purified from calf thymus using immunoaffinity column chromatography as described by Tamai et al. [39]. Recombinant rat pol β was purified from *Escherichia coli* JMpβ5 as described by Date et al. [40]. The human pol γ catalytic gene was cloned into the pFastBac vector. A histidine-tagged enzyme was expressed using the BAC-TO-BAC HT Baculovirus Expression System according to supplier instructions (Life Technologies, Inc., Gaithersburg, MD, USA) and purified using ProBoundresin (Invitrogen, Carlsbad, CA, USA) [41]. Human pols δ and ε were purified by nuclear fractionation of human peripheral blood cancer cells (Molt-4) by carrying out affinity column chromatography of the second subunits of pols δ and ε, respectively [42]. A truncated form of human pol η (residues 1–511) tagged with His_6 at its C-terminus was expressed in *E. coli* cells and purified by the method of Kusumoto et al. [43]. A recombinant mouse pol ι tagged with His_6 at its C-terminus was expressed and purified by undertaking nickel–nitrilotriacetic acid column chromatography, as described previously [44]. A truncated form of pol κ (residues 1–560) with His_6-tags attached at the C-terminus was overproduced in *E. coli* and purified according to the method of Ohashi et al. [45]. Recombinant human His-pol λ was overexpressed and purified according to the method of Shimazaki et al. [46]. Pol α from a higher plant, cauliflower inflorescence, was purified according to the method of Sakaguchi et al. [47]. Calf TdT, *Taq* pol and T4 pol were purchased from Takara Bio, Inc. (Shiga, Japan). The Klenow fragment of pol I from *E. coli* was purchased from Worthington Biochemical Corp. (Lakewood, NJ, USA).

3.2. Pol Assays

The reaction mixtures for calf pol α, rat pol β, plant pol α and prokaryotic pols have been described [48, 49]. Those for human pol γ and as well as for human pols δ and ε were as described by Umeda et al. [41] and Ogawa et al. [50], respectively. Reaction mixtures for mammalian pols η, ι and κ were the same as that for calf pol α, and the reaction mixture for human pol λ was the same as that for rat pol β. Poly(dA)/oligo$(dT)_{18}$ (A/T = 2/1) and tritium-labeled 2′-deoxythymidine 5′-triphosphate ($[^3H]$-dTTP) were used as the DNA template-primer substrate and nucleotide (i.e., 2′-deoxyribonucleoside 5′-triphosphate [dNTP]) substrate, respectively). For TdT reactions, oligo$(dT)_{18}$ (3′-OH) and dTTP were used as the DNA primer substrate and dNTP substrates, respectively.

Flavonoids were dissolved in distilled dimethyl sulfoxide (DMSO) to various concentrations and sonicated for 30 s. Aliquots (4 μL) of sonicated samples were mixed with 16 μL of each enzyme (final amount, 0.05 U) in 50 mM Tris-HCl (pH 7.5) containing 1 mM dithiothreitol, 50% glycerol (by vol), and 0.1 mM ethylenediamine tetraacetic acid (EDTA), and incubated at 0°C for 10 min. Next. these inhibitor–enzyme mixtures in 8-μL volumes were added to 16 μL of each of the enzyme standard reaction mixtures and incubated at 37°C for 60 min except for *Taq* pol, which was incubated at 74°C for 60 min. Activity without the inhibitor was considered to be 100%, and the relative activity determined for each inhibitor concentration. One unit of pol activity was defined as the amount of each enzyme that catalyzed incorporation of 1 nmol dNTP (i.e., dTTP) into synthetic DNA template-primers in 60 min at 37°C, and under normal reaction conditions [48, 49].

3.3. Pol Inhibition by Flavonoids

Herein, ten plant flavonoids were selected and prepared for further study. The flavonoids were chalcone (**1**), resveratrol (**2**) and three major classes: flavanones (**3**–**5**), flavones (**6**–**8**) and isoflavones (**9** and **10**) (Figure 1). These flavonoids were purchased from Sigma–Aldrich Inc. (Saint Louis, MO, USA), and the purity of these flavonoids was ≈98%.

The inhibitory activity of each flavonoid toward mammalian pols was investigated using calf pol α and human pols κ and λ. Mammalian pols α, κ and λ were chosen as a representative replicative pol (B-family pols), a translation synthesis (TLS) repair pol (Y-family pols) and a repair/recombination-related pol (X-family pols), respectively [29–32]. Assessment of the relative activity of each pol at a set concentration (100 μM) of the ten test compounds showed that inhibition of the relative activity of these three pols by <80% was observed for naringenin (**4**), hesperetin (**5**), apigenin (**7**), daidzein (**9**) and genistein (**10**) (Figure 2).

Apigenin (**7**) exhibited the strongest inhibitory effect on mammalian pols of all the tested flavonoids. Apigenin-7-*O*-glucoside (**8**) showed no pol inhibitory activities, suggesting that the aglycone structure (lacking a sugar group) of flavonoids must be important for inhibition.

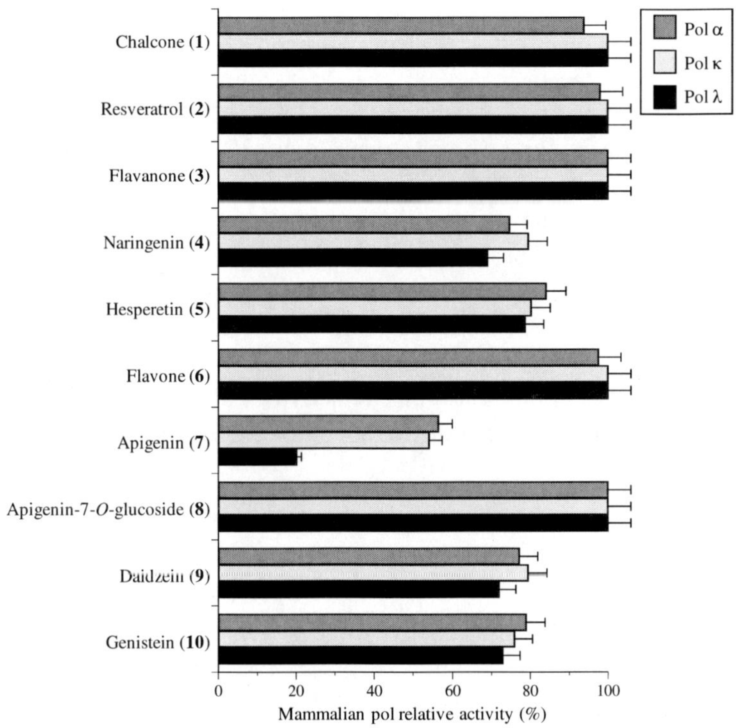

Figure 2. Effect of plant flavonoids (**1–10**) on the activity of mammalian pols. Each compound at 100 μM was incubated with calf pol α (B-family pol), human pol κ (Y-family pol) and human pol λ (X-family pol) (0.05 units each). Percent relative activity is shown. Pol activity in the absence of the compound was considered to be 100 %. Data are the mean ± S.D. of three independent experiments.

These results suggested that the trend in pol inhibitory activity may be related to the numbers of hydroxyl groups in the flavonoid backbone. Therefore, three hydroxyl groups in the molecule might be important flavonoid structural characteristics for pol inhibition.

These compounds inhibited the activities of pols α, κ and λ, with the effects on pol λ being slightly stronger than those on pols α and κ. By contrast, the other flavonoids (**1–3**, **6** and **8**) had no influence on the activities of these pols. When activated DNA (bovine deoxyribonuclease I-treated DNA) was used as the DNA template-primer substrate instead of synthesized DNA [poly(dA)/oligo$(dT)_{18}$ (A/T = 2/1)], and dNTP was used as the nucleotide substrate instead of dTTP, the inhibitory effects of these compounds did not change (data not shown).

4. Effects of Apigenin on the Activities of Various Species of Pols

4.1. Inhibition of Various Pols by Apigenin

Among the ten major flavonoids investigated, apigenin displayed the strongest inhibitory effect on human pol λ (Figure 2) and is, therefore, the focus of this section. Briefly, we obtained 11 mammalian pol species: α, β, γ, δ, ε, η, ι, κ, λ, μ and TdT. Table 1 shows the inhibitory effect (half-maximal inhibitory concentration (IC_{50}) value) of apigenin against various pol species, including the mammalian pols obtained. This compound inhibited the activities of all 11 mammalian pols tested from four families. Pol λ and pols δ and ι showed the strongest and weakest inhibition, respectively, with IC_{50} values of 42.5 and 125 μM, respectively. Interestingly, the inhibitory effect of apigenin on X-family pols was stronger than that on A-, B- and Y-families of pols. In contrast, apigenin showed no inhibitory effect on plant (i.e., cauliflower) pol α, or prokaryotic pols, such as the Klenow fragment of *E. coli* pol I, *Taq* pol and T4 pol. These results suggest that apigenin probably selectively inhibits mammalian pols, particularly the X-family of pol λ.

Table 1. IC_{50} of apigenin (7) on the activities of mammalian pols, pols of other species, and various DNA metabolic enzymes

Enzyme	IC_{50} (μM)
(1) Mammalian pols	
[A-Family]	
Human pol γ	101 ± 5.2
[B-Family]	
Calf pol α	116 ± 5.9
Human pol δ	125 ± 6.2
Human pol ε	121 ± 6.1
[X-Family]	
Rat pol β	77.7 ± 3.9
Human pol λ	42.5 ± 2.2
Human pol μ	53.6 ± 2.8
Calf TdT	63.7 ± 3.3
Enzyme	IC_{50} (μM)
[Y-Family]	
Human pol η	119 ± 6.0
Mouse pol ι	125 ± 6.3
Human pol κ	109 ± 5.6
(2) Other pols	
Cauliflower pol α	> 200
Escherichia coli pol I	> 200
Taq pol	> 200
T4 pol	> 200
(3) Other DNA metabolic enzymes	
Mouse IMP dehydrogenase (type II)	> 200
T4 polynucleotide kinase	> 200
Bovine deoxyribonuclease I	> 200

Apigenin (**7**) was incubated with each enzyme (0.05 units) and other DNA metabolic enzymes. One unit of pol activity was defined as the amount of enzyme catalyzing the incorporation of 1 nmol of dNTP substrate (i.e., dTTP) into the synthetic DNA template-primer substrate (i.e., poly(dA)/oligo$(dT)_{18}$, A/T = 2/1) in 60 min at 37°C under normal reaction conditions for each enzyme. Enzyme activity in the absence of an inhibitor (control) was taken to be 100%. Data are the mean ± S.D. of three independent experiments.

4.2. Mode of Inhibition of Mammalian Pols α, κ and λ by Apigenin

Next, to elucidate the mechanism of selective inhibition of mammalian pols by apigenin, we investigated the mode of inhibition of this compound against pols α, κ and λ. For kinetic analyses, poly(dA)/oligo$(dT)_{18}$ and dTTP were used as the synthetic DNA template-primer substrate and nucleotide (dNTP) substrate, respectively. The extent of inhibition as a function of DNA template-primer substrate or dNTP substrate concentration was measured (Table 2).

Table 2. Kinetic analyses of the inhibitory effects of apigenin (7) on the activity of mammalian pols α, κ and λ as a function of the DNA template-primer dose and nucleotide substrate concentration

Pol	Substrate	Apigenin (μM)	K_m [a)] (μM)	V_{max} [a)] (pmol/h)	K_i [b)] (μM)	Inhibition type
α	DNA template-primer [c)]	0	13.0	55.6	63.6	Noncompetitive
		30		33.3		
		60		23.8		
		90		18.5		
	dNTP substrate [d)]	0	1.65	29.2	82.4	Noncompetitive
		30		16.9		
		60		11.9		
		90		9.1		
κ	DNA template-primer [c)]	0	1.72	40.0	57.8	Noncompetitive
		30		23.8		
		60		16.9		
		90		13.2		
	dNTP substrate [d)]	0	2.00	41.6	76.1	Noncompetitive
		30		22.2		
		60		15.2		
		90		11.5		
λ	DNA template-primer [c)]	0	2.38	83.3	24.5	Noncompetitive
		10		52.6		
		20		37.0		
		30		28.6		
	dNTP substrate [d)]	0	1.18	52.6	19.8	Competitive
		10	1.56			
		20	2.33			
		30	4.55			

[a] Data were obtained from Lineweaver–Burk plots.

[b] Data were obtained from Dixon plots.

[c] poly(dA)/oligo$(dT)_{18}$.

[d] dTTP.

Double reciprocal (i.e., Lineweaver–Burk) plots showed that apigenin-induced inhibition of calf pol α activity was noncompetitive with respect to the DNA template-primer substrate and dNTP substrate. For the DNA template-primer substrate, the apparent Michaelis constant (K_m) was unchanged at 13.0 μM, whereas decreases of 55.6, 33.3, 23.8 and 18.5 pmol/h in maximum velocity (V_{max}) were observed in the presence of 0, 30, 60 and 90 μM apigenin, respectively. K_m for the dNTP substrate was unchanged at 1.65 μM, and V_{max} for the dNTP substrate decreased from 29.2 to 9.1 pmol/h in the presence of 90 μM apigenin. Inhibition constant (K_i) values, obtained based on Dixon plots, were found to be 63.6 and 82.4 μM for the DNA template-primer substrate and dNTP substrate, respectively.

Data were expressed as double reciprocal plots, which showed that apigenin inhibited human pol κ activity in a noncompetitive manner with respect to the DNA template-primer substrate and dNTP substrate. For the DNA template-primer substrate, the apparent K_m was unchanged at 1.72 μM apigenin, whereas decreases in V_{max} of 42.1, 59.2 and 68.8% were observed in the presence of apigenin at 30, 60 and 90 μM, respectively. K_m for the dNTP substrate was unchanged at 2.00 μM, and V_{max} for the dNTP substrate decreased from 41.6 to 11.5 pmol/h in the presence of 90 μM apigenin. K_i values, obtained from Dixon plots, were 57.8 μM for the DNA template-primer substrate and 76.1 μM for the dNTP substrate.

Apigenin-induced inhibition of pol λ activity was noncompetitive and competitive with respect to the DNA template-primer substrate and dNTP substrate, respectively. For the DNA template-primer, the apparent K_m was unchanged at 2.38 μM, whereas decreases in V_{max} of 83.3–28.6 pmol/h were observed in the presence of 30 μM apigenin. The V_{max} for the dNTP substrate was unchanged at 52.6 pmol/h, and the K_m for the dNTP substrate increased from 1.18 to 4.55 μM in the presence of 30 μM apigenin. K_i values were 24.5 and 19.8 μM for the DNA template-primer substrate and dNTP substrate, respectively. The K_i value for the dNTP substrate was ≈1.2-fold smaller than that for the DNA template-primer substrate, so the affinity of apigenin was greater for the enzyme-dNTP substrate binary complex than for the enzyme-DNA template-primer substrate complex. The type of competitive inhibition of apigenin for pol λ might be important for the strongest inhibitory activity against X-family pols among the mammalian pol families (Table 1).

When activated DNA and 4 dNTPs were used as DNA template-primer substrate and nucleotide substrates, respectively, the mode of inhibition of these mammalian pols by apigenin was the same as that observed for the synthetic DNA template-primer (data not shown).

These results suggest that apigenin binds to or interacts with a site distinct from the DNA template-primer substrate-binding site and dNTP substrate-binding site of each pol.

5. Effects of Apigenin on the Activities of Other DNA Metabolic Enzymes

5.1. Other DNA Metabolic Enzyme Assays

Mouse inosine-5'-monophosphate (IMP) dehydrogenase (type II), T4 polynucleotide kinase and bovine deoxyribonuclease I were prepared from commercial sources. The activities of these enzymes were determined using standard assays according to manufacturer specifications and as per the method of Mizushina et al. [51], Soltis and Uhlenbeck [52] and Lu and Sakaguchi [53], respectively.

5.2. Inhibition of Other DNA Metabolic Enzymes by Apigenin

Apigenin influenced the activities of other DNA-metabolic enzymes such as mouse IMP dehydrogenase (type II), T4 polynucleotide kinase and bovine deoxyribonuclease I only minimally (Table 1). These results suggest that apigenin selectively inhibits the activities of mammalian pols among the DNA metabolic enzymes tested.

Interestingly, some flavonoids, particularly apigenin, are inhibitors of mammalian pols.

Next, we undertook specific assays to determine whether apigenin-induced inhibition resulted from the ability of the compound to bind to DNA or to the enzyme. The interaction of apigenin with double-stranded DNA (dsDNA) was investigated by examining its thermal transition. For this, the melting temperature (T_m) of dsDNA in the presence of excess apigenin (200 μM) was observed using a spectrophotometer (260 nm) equipped with a thermoelectric cell holder. Thermal transition (i.e., T_m) from 75°C to 90°C was not observed within the concentration range used in the assay (Figure 3). The typical intercalating compound, ethidium bromide, (EtBr, 15 μM) was used as a positive control, and a clear thermal transition was observed.

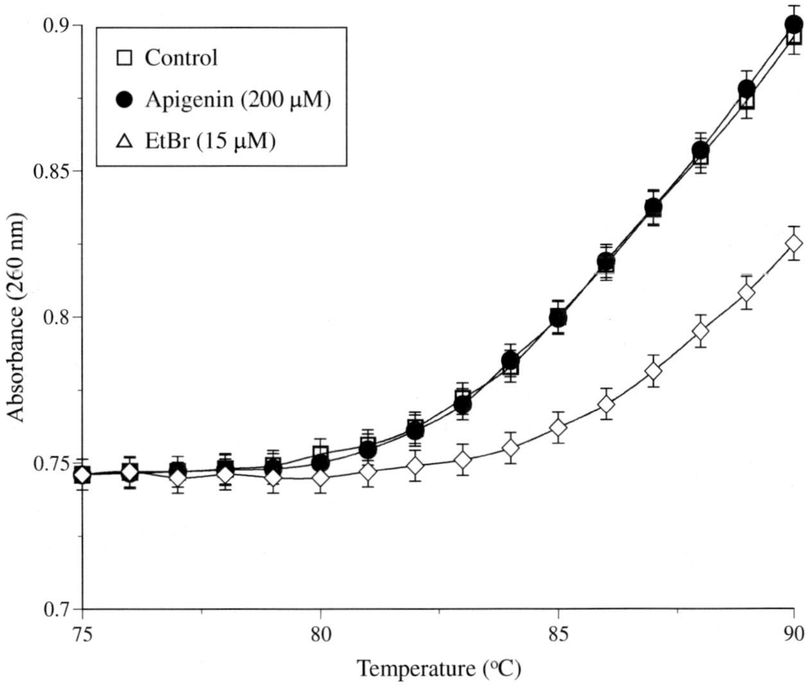

Figure 3. Effect of apigenin (**7**) on the thermal transition of dsDNA. Control (open squares), apigenin (200 μM, closed circles) and ethidium bromide (EtBr, 15 μM, open diamonds) were incubated with 6 μg/mL of calf thymus dsDNA in 0.1 M Na-phosphate buffer (pH 7.0). Data are the mean ± S.D. of three independent experiments.

Whether the inhibitory effect of apigenin resulted from nonspecific adhesion to mammalian pol molecules or from its selective binding to specific sites was investigated by determining if an excessive amount of nucleic acid (poly[rC]) or protein (bovine serum albumin (BSA)) could prevent the inhibitory effect of apigenin. Poly(rC) and BSA had little or no influence on pol inhibition by apigenin (data not shown), suggesting that this compound selectively binds to mammalian pol molecules. These observations suggest that apigenin does not act as a DNA-intercalating agent or as a DNA template-primer substrate but that this compound binds directly to the enzyme to inhibit its activity.

Collectively, these results suggest that apigenin is a potent and specific inhibitor of mammalian pols. We investigated in detail whether pol inhibition by major flavonoids in plants was effective for proliferation of human cancer cells.

6. Effects of Flavonoids on Growth of Human Cancer Cells

6.1. Cell Culture and Measurement of the Viability of Cancer Cells

The HCT116 human colon carcinoma cell line was obtained from the American Type Culture Collection (Manassas, VA, USA). Cells were cultured in RPMI 1640 medium supplemented with 10% fetal bovine serum, penicillin (100 units/mL), streptomycin (100 μg/mL), and 1.6 mg/mL $NaHCO_3$ at 37°C in a humid atmosphere of 5% CO_2/95% air. For the cell viability assay, cells were seeded at 1×10^3 cells/well in a 96-well microplate with various concentrations of flavonoids, and incubated for 48 h. Cell viability was determined by the WST-1 assay [54].

6.2. Suppression of the Growth of Human Cancer Cells by Flavonoids

Pols have emerged as important cellular targets for chemical intervention in the development of anti-cancer agents [55]. Therefore, flavonoids, which have pol inhibitory activity, could be useful in chemotherapy. Hence, we investigated the cytotoxic effect of these compounds against human colon carcinoma (HCT116) cells. Among 100 μM of the ten flavonoids, naringenin (**4**), hesperetin (**5**), apigenin (**7**), daidzein (**9**) and genistein (**10**) showed the <80% of cell viability (Figure 4). Apigenin (**7**) had the strongest inhibitory effect on the proliferation of HCT116 cells among the compounds tested, and at 10 and 100 μM of apigenin showed ≈20% and ≈85% of cell viability, respectively. Conversely, 100 μM of the other flavonoids (**1–3**, **6** and **8**) did not suppress cell proliferation. The effect of 100 μM of flavonoids on the viability of HCT116 cells showed a very high correlation with the effect of 100 μM of these compounds on the relative activity of mammalian pols (Figure 2). This finding suggested that suppression of cell proliferation by flavonoids might be based on pol inhibition.

Of all the flavonoids tested, apigenin (**7**) exhibited the strongest effect on inhibition of mammalian pols α, κ and λ (Figure 2) and suppression of the growth of HCT116 cells (Figure 4). Therefore, we concentrated our efforts on apigenin in the latter part of this study.

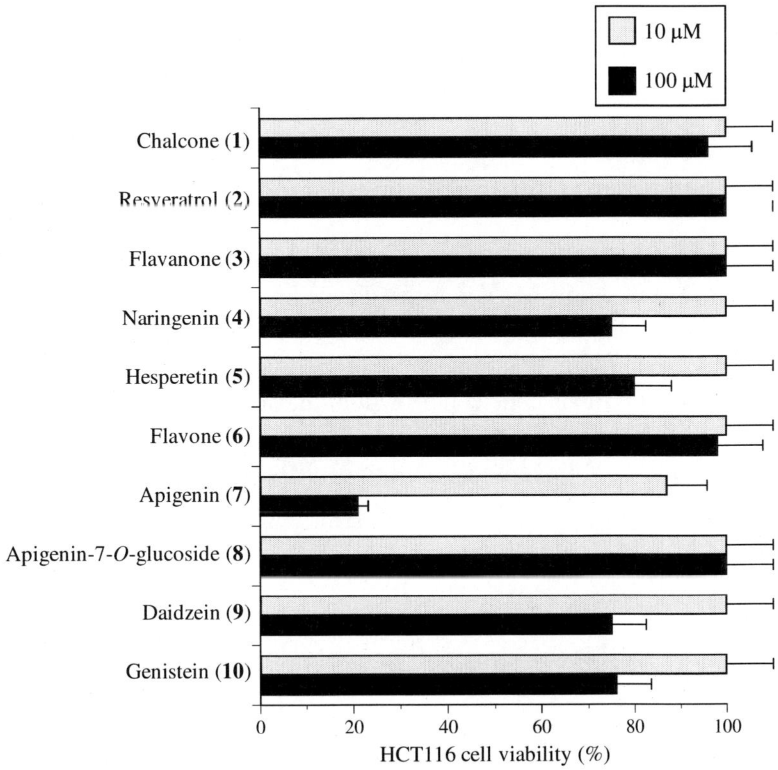

Figure 4. Effect of plant flavonoids (**1–10**) on the proliferation of the growth of HCT116 human colon carcinoma cultured cells. Each compound (10 and 100 μM) was added to the culture of HCT116 cells. Cells were incubated for 24 h, and the rate of inhibition of growth of cultured cells was determined by the WST-1 assay [54]. Inhibition of growth of the cancer cells in the absence of the compound was taken to be 100%. Data are the mean ± S.D. of five independent experiments.

7. Effect of Apigenin on Cultured Human Cancer Cells and Normal Cells

7.1. Cell Culture and Measurement of Cell Viability and Apoptotic Cells

The following human cancer cell lines were obtained from the American Type Culture Collection: lung (A549), prostate (PC3), colon (HCT116), cervix (HeLa), hepatocellular liver (HepG2) and breast (MCF-7). The normal cell line of human dermal fibroblasts, HDF (CS-2FO-101), was purchased from Dainippon Sumitomo Pharma Co., Ltd. (Osaka, Japan). These cells were cultured with apigenin for 48 h, and cell viability measured by the WST-1 assay [54]. Aliquots of HCT116 cells (2.5×10^4) were plated in each well of an eight-well chamber slide (Nunc, Rochester, NY, USA), and cells were incubated with apigenin for 48 h at 37°C. The terminal deoxynucleotidyl transferase dUTP nick end labeling (TUNEL) assay is a method for detecting DNA fragmentation by labeling the terminal end of nucleic acids. The percentage of apoptotic cells was determined using the ApopTag Red *in situ* apoptosis detection kit (Chemicon, Temecula, CA, USA). Culture dishes were stained, and the percentage of apoptotic cells examined under an IX70 fluorescence microscope (Olympus, Tokyo, Japan).

7.2. Effect of Apigenin on Cultured Human Cancer Cells and Normal Cells

Apigenin treatment for 48 h dose-dependently suppressed HCT116 cell growth, with a 50% lethal dose (LD_{50}) of 34.4 μM (Figure 5). This LD_{50} was similar to the IC_{50}s for pols, suggesting that apigenin might be able to penetrate the cell membrane and reach the nucleus, where it may be able to inhibit the activities of pols, leading to suppression of cell growth. By contrast, apigenin did not influence the cell growth of the normal cell line, HDF.

Table 3. LD_{50} values of apigenin (7) on the proliferation of human cancer cells

Human cancer cell line	LD_{50} (μM)
A549 (lung)	35.4 ± 3.7
HCT116 (large intestine)	34.4 ± 3.6
HeLa (cervical)	41.0 ± 4.5
HepG2 (hepatocellular liver)	37.4 ± 3.8
MCF-7 (breast)	37.2 ± 3.7
PC3 (prostate gland)	38.8 ± 3.9

The six human cancer cell lines were incubated with apigenin (**7**) for 48 h. Cell viability was determined by the WST-1 assay [54], and this viability in the absence of an inhibitor (control) was taken to be 100%. Data are the mean ± S.D. of five independent experiments.

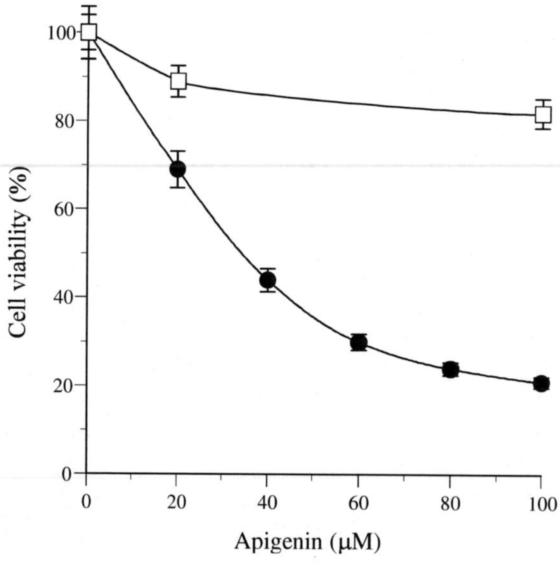

Figure 5. Effect of apigenin (**7**) on the growth of human cancer cells and normal cultured cells. Dose–response curve of the survival of human colon carcinoma HCT116 cells (closed circle) and human dermal fibroblasts (HDF) (open square). Cells were cultured for 48 h in media containing the indicated concentrations of apigenin. Cell proliferation was determined by the WST-1 assay [54]. Inhibition of cell growth in the absence of the compound was taken to be 100%. Values are the mean ± S.D. for five independent experiments.

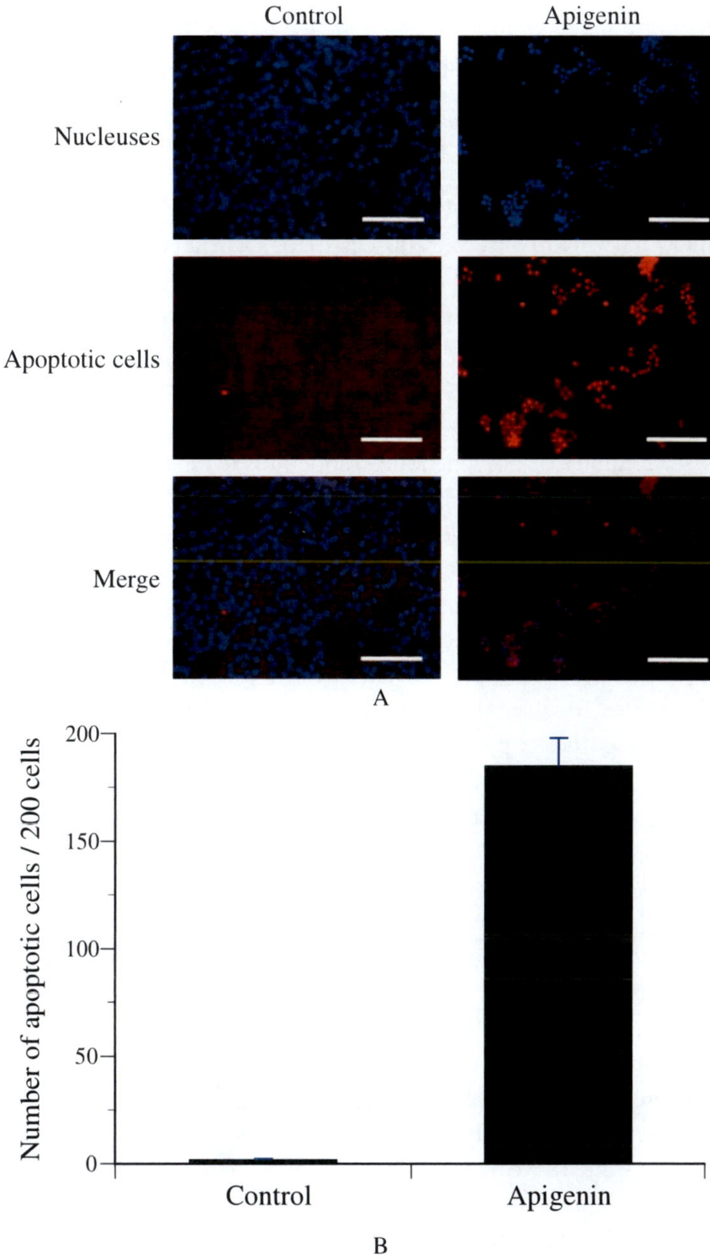

Figure 6. Apoptotic effect of apigenin (**7**) on HCT116 human colon carcinoma cultured cells.

Apigenin prevented the proliferation of six human cancer cell lines (A549, HCT116, HeLa, HepG2, MCF-7, PC3), with LD_{50} values of 34.4–41.0 μM (Table 3). These results suggested that this compound could have suppressive activity against different cancer cell lines. In particular, this compound showed the strongest suppression of cell proliferation in the colon cancer cell line, HCT116.

Apoptosis is programmed cell death that occurs in multicellular organisms [56]. Biochemical events can trigger characteristic morphologic changes and cell death. One characteristic feature of apoptotic cells is marked elevation in levels of DNA fragmentation. Apigenin significantly induced apoptosis as detected by the TUNEL assay (Figure 6). Given that inducers of apoptosis often display anti-cancer activity, this compound could be a potential anti-cancer chemotherapy agent.

(A) HCT116 cells were treated without (control, left panels) or with 100 μM apigenin (**7**) (right panels). Bars represent 100 μm. (B) The number of apoptotic cells induced by apigenin (**7**). Cells were incubated for 48 h with DMSO (control) or 100 μM apigenin (**7**). Apoptotic cells were counted individually from ≥200 cells (for each condition). Data are the mean ± S.D. for three independent experiments

8. Discussion

Flavonoids are a family of polyphenolic compounds synthesized by plants with a similar structure. They are divided into subclasses, including flavanones, flavones and isoflavones [57]. Several beneficial properties have been attributed to these dietary compounds, including anti-oxidant, anti-inflammatory and anti-carcinogenic effects. According to estimated values, average intakes of flavonoids have ranged from 6 mg/day in Finland to 64 mg/day in Japan, with intermediate intakes in the USA (13 mg/day), Italy (27 mg/day) and the Netherlands (33 mg/day). These estimates were based on analyses of plant flavonoids such as apigenin in composite food samples for the population analyzed in the Seven Countries Study [58]. In another study in a Hungarian population, the intake of flavonoids was lower than in citizens from the Netherlands (23 mg/day), Denmark (28 mg/day) and Finland (55 mg/day). Intake of flavonoids in 17 diets was estimated [59]. The lowest intake (1–9 mg/day) was estimated in South African diets, whereas the highest flavonoid intake (75–81 mg/day) was from a Scandinavian diet. In addition to flavonoid intake, dietary sources of the flavones vary among countries, with major contributions from tea in Japan (95%) and the Netherlands (64%), red wine and beer in Italy (46%) and vegetables and fruits in Finland (100%) and the USA (80%). In Australia, tea remains the major dietary flavonoid, with apparent dietary consumption of ≤351 mg/person/day, of which 75% are flavan-3-ols [60]. One of the molecules that has garnered considerable interest as a beneficial agent for human health with cancer-preventive and/or therapeutic properties is the plant flavone apigenin.

Apigenin is present abundantly in: common fruits (e.g., oranges, grapefruit); plant-derived beverages (e.g., tea); vegetables (e.g., parsley, onions); chamomile; wheat sprouts; and some seasonings. One of the most common sources of apigenin consumed as a single-ingredient herbal tea is chamomile, prepared from the dried flowers from *Matricaria chamomilla* [61]. This is an annual herbaceous plant indigenous to Europe and Western Asia that has been naturalized in Australia, Britain and the USA. Also known as "German chamomile", "Hungarian chamomile", "mayweed", "sweet false chamomile" or "wild chamomile", the plant is cultivated in Germany, Hungary, Russia and other Southern and Eastern European countries for its flower heads. Infusions of chamomile contain maximum concentrations of apigenin (0.8–1.2%) and essential oils that have aromatic, flavoring and coloring properties. Chamomile is consumed in the form of tea at >1 million cups per day. Other sources of apigenin include beverages such as wine and beer brewed from natural ingredients. Apigenin is commonly present as a constituent in red wine [62]. Like red wine, beer also provides a good source of apigenin [63].

Apigenin has garnered particular interest in recent years as a beneficial and health-promoting agent because of its low intrinsic toxicity and because of its striking effects on normal *versus* cancerous cells compared with other structurally related flavonoids [64]. There is very little evidence to suggest that apigenin promotes adverse metabolic reactions *in vivo* if consumed as part of a normal diet. In recent years, apigenin has been recognized increasingly as a cancer-chemopreventive agent. Interest in the possible cancer-preventive effects of apigenin has increased owing to reports of potent anti-oxidant and anti-inflammatory activities. Indirect support for this hypothesis is correlated with a study in which consumption of flavonoid-free diets by healthy human volunteers was reported to lead to a decrease in markers of oxidative stress in blood (i.e., anti-oxidant vitamins, erythrocyte superoxide dismutase (SOD) activity and lymphocyte DNA damage in plasma) commonly associated with an enhanced risk of disease, suggesting the beneficial effects of flavonoids [65].

In this review, it was found that apigenin showed the strongest inhibitor of the activities of mammalian pols α, κ and λ among the plant flavonoids investigated (Figure 2). Its cytotoxicity in human colon carcinoma (HCT116) cells (Figures 4 and 5) and induction of apoptosis (Figure 6) was realized through selective inhibition of pols (Table 1), which are essential for the replication, repair and recombination of DNA as well as for cell division. Apigenin also suppressed the other human cancer cell lines, but did not influence the cell growth of the normal cell line (Table 3). Thus, apigenin could possess selective cytotoxic effects against cancer cells without side effects. These results suggest a novel mechanism of action of the cancer chemopreventive effect of apigenin based on the inhibition of mammalian pols. Therefore, the chemically synthesized apigenin and the apigenin-enriched fraction in plant could potentially be used in drugs and functional foods for cancer prevention, respectively.

ACKNOWLEDGMENTS

This study was supported in part by the Ministry of Education, Culture, Sports, Science and Technology (MEXT), Japan-Supported Program for the Strategic Research Foundation at Private Universities, 2012–2016. Y.M. received Grant-in-Aids for Scientific Research (C) (No. 24580205) from MEXT, Japan and Cosmetology Research Foundation (25th, 2014, Japan).

REFERENCES

[1] World Health Organization: Cancer, 2006.

[2] Jemal, A., Murray, T., Ward, E., Samuels, A., Tiwari, R.C., Ghafoor, A., Feuer, E.J. and Thun, M.J. (2005). National Cancer Institute and mortality data from the National Center for Health Statistics. Cancer statistics, 2005. *CA Cancer J. Clin.*, *55*, 10-30.

[3] American Cancer Society. http://www.cancer.org/ 2006.

[4] Jhanwar, Y.S. and Divgi, C. (2005). Current status of therapy of solid tumors. *J. Nucl. Med.*, *46*, 141S-150S.
[5] Abou-Jawde, R., Choueiri, T., Alemany, C. and Mekhail, T. (2003). An overview of targeted treatments in cancer. *Clin. Ther.*, *25*, 2121-2137.
[6] Guillemard, V. and Saragovi, H.U. (2004). Novel approaches for targeted cancer therapy. *Curr. Cancer Drug Targets*, *4*, 313-326.
[7] Antonia, S., Mule, J.J. and Weber, J.S. (2004). Current developments of immunotherapy in the clinic. *Curr. Opin. Immunol.*, *16*, 130-136.
[8] Caponigro, F., Basile, M., De Rosa, V. and Normanno, N. (2005). New drugs in cancer therapy. National Tumor Institute, Naples, 2004. *Anticancer Drugs*, *16*, 211-221.
[9] Miner, T.J. (2005). Palliative surgery for advanced cancer: lessons learned in patient selection and outcome assessment. *Am. J. Clin. Oncol.*, *28*, 411-414.
[10] Elshaikh, M., Ljungman, M., Ten Haken, R. and Lichter, A.S. (2006). Advances in Radiation Oncology. *Annu. Rev. Med.*, *57*, 19-31.
[11] Smith, J.J., Tully, P. and Padberg, R.M. (2005). Chemoprevention: a primary cancer prevention strategy. *Semin. Oncol. Nurs.*, *21*, 243-251.
[12] Sporn, M.B. and Liby, K.T. (2005). Cancer chemoprevention: scientific promise, clinical uncertainty. *Nat. Clin. Pract. Oncol.*, *2*, 518-525.
[13] Brenner, D.E. and Gescher, A.J. (2005). Cancer chemoprevention: lessons learned and future directions. *Br. J. Cancer*, *93*, 735-739.
[14] Kim, E.S. and Hong, W.K. (2005). An apple a day...does it really keep the doctor away? The current state of cancer chemoprevention. *J. Natl. Cancer Inst.*, *97*, 468-470.
[15] Shukla, S. and Gupta, S. (2005). Dietary agents in the chemoprevention of prostate cancer. *Nutr. Cancer*, *53*, 18-32.
[16] Crowell, J.A. (2005). The chemopreventive agent development research program in the Division of Cancer Prevention of the US National Cancer Institute: an overview. *Eur. J. Cancer*, *41*, 1889-1910.
[17] Birt, D.F., Hendrich, S. and Wang, W. (2001). Dietary agents in cancer prevention: flavonoids and isoflavonoids. *Pharmacol. Ther.*, *90*, 157-177.
[18] Kelloff, G.J., Crowell, J.A., Steele, V.E., Lubet, R.A., Malone, W.A., Boone, C.W., Kopelovich, L., Hawk, E.T., Lieberman, R., Lawrence, J.A., Ali, I., Viner, J.L. and Sigman, C.C. (2000). Progress in cancer chemoprevention: development of diet-derived chemopreventive agents. *J. Nutr.*, *130*. 467S-471S.
[19] Genkinger, J.M., Platz, E.A., Hoffman, S.C., Comstock, G.W. and Helzlsouer, K.J. (2004). Fruit, vegetable, and antioxidant intake and all-cause, cancer, and cardiovascular disease mortality in a community-dwelling population in Washington County, MD. *Am. J. Epidemiol.*, *160*, 1223-1233.
[20] Cohen, J.H., Kristal, A.R. and Stanford, J.L. (2000). Fruit and vegetable intakes and prostate cancer risk. *J. Natl. Cancer Inst.*, *92*, 61-68.
[21] Terry, P., Giovannucci, E., Michels, K.B., Bergkvist, L., Hansen, H., Holmberg, L. and Wolk, A. (2001). Fruit, vegetables, dietary fiber, and risk of colorectal cancer. *J. Natl. Cancer Inst.*, *93*, 525-533.
[22] Le Marchand, L. (2002). Cancer preventive effects of flavonoids – a review. *Biomed. Pharmacother.*, *56*, 296-301.
[23] Surh, Y.J. (2003). Cancer chemoprevention with dietary phytochemicals. *Nat. Rev. Cancer*, *3*, 768-780.

[24] Manach, C., Scalbert, A., Morand, C., Remesy, C. and Jimenez, L. (2004). Polyphenols: food sources and bioavailability. *Am. J. Clin. Nutr.*, *79*, 727-747.

[25] Yang, C.S., Landau, J.M., Huang, M.T. and Newmark, H.L. (2001). Inhibition of carcinogenesis by dietary polyphenolic compounds. *Annu. Rev. Nutr.*, *21*, 381-406.

[26] O'Prey, J., Brown, J., Fleming, J. and Harrison, P.R. (2003). Effects of dietary flavonoids on major signal transduction pathways in human epithelial cells. *Biochem. Pharmacol.*, *66*, 2075-2088.

[27] Thiery-Vuillemin, A., Nguyen, T., Pivot, X., Spano, J.P., Dufresnne, A. and Soria, J.C. (2005). Molecularly targeted agents: their promise as cancer chemopreventive interventions. *Eur. J. Cancer*, *41*, 2003-2015.

[28] Kornberg, A. and Baker, T.A. (1992). Eukaryotic DNA polymerase, "DNA replication" 2nd edition. Freeman, W.D., and Co., New York, Chapter 6, pp. 197-225.

[29] Hübscher, U., Maga, G. and Spadari, S. (2002). Eukaryotic DNA polymerases. *Annu. Rev. Biochem.*, *71*, 133-163.

[30] Bebenek, K. and Kunkel, T.A. (2004). DNA Repair and Replication, "Advances in Protein Chemistry", Elsevier, San Diego, Yang, W. (Ed.), vol. 69, pp. 137-165.

[31] Takata, K., Shimizu, T., Iwai, S. and Wood, R.D. (2006). Human DNA polymerase N (POLN) is a low fidelity enzyme capable of error-free bypass of 5S-thymine glycol. *J. Biol. Chem.*, *281*, 23445-23455.

[32] Lange, S.S., Takata, K. and Wood, R.D. (2011). DNA polymerases and cancer. *Nat. Rev. Cancer*, *11*, 96-110.

[33] Loeb, L.A. and Monnat, Jr. R.J. (2008). DNA polymerases and human disease. *Nat. Rev. Genet.*, *9*, 594-604.

[34] Mizushina, Y., Yagi, H., Tanaka, N., Kurosawa, T., Seto, H., Katsumi, K., Onoue, M., Ishida, H., Iseki, A., Nara, T., Morohashi, K., Horie, T., Onomura, Y., Narusawa, M., Aoyagi, N., Takami, K., Yamaoka, M., Inoue, Y., Matsukage, A., Yoshida, S. and Sakaguchi, K. (1996). Screening of inhibitor of eukaryotic DNA polymerases produced by microorganisms. *J. Antibiot. (Tokyo)*, *49*, 491-492.

[35] Mizushina, Y. (2009). Specific inhibitors of mammalian DNA polymerase species. *Biosci. Biotechnol. Biochem.*, *73*, 1239-1251.

[36] Mizushina, Y. (2011). Screening of novel bioactive compounds from food components and nutrients. *J. Jpn. Soc. Nutr. Food Sci.*, *64*, 377-384.

[37] Mizushina, Y., Ishidoh, T., Kamisuki, S., Nakazawa, S., Takemura, M., Sugawara, F., Yoshida, H., Sakaguchi, K. (2003). Flavonoid glycoside: a new inhibitor of eukaryotic DNA polymerase alpha and a new carrier for inhibitor-affinity chromatography. *Biochem. Biophys. Res. Commun.*, *301*, 480-487.

[38] Mizushina, Y., Saito, A., Tanaka, A., Nakajima, N., Kuriyama, I., Takemura, M., Takeuchi, T., Sugawara, F., Yoshida, H. (2005). Structural analysis of catechin derivatives as mammalian DNA polymerase inhibitors. *Biochem. Biophys. Res. Commun.*, *333*, 101-109.

[39] Tamai, K., Kojima, K., Hanaichi, T., Masaki, S., Suzuki, M., Umekawa, H. and Yoshida, S. (1998). Structural study of immunoaffinity-purified DNA polymerase α-DNA primase complex from calf thymus. *Biochim. Biophys. Acta*, *950*, 263-273.

[40] Date, T., Yamaguchi, M., Hirose, F., Nishimoto, Y., Tanihara, K. and Matsukage, A. (1998). Expression of active rat DNA polymerase β in *Escherichia coli*. *Biochemistry*, *27*, 2983-2990.

[41] Umeda, S., Muta, T., Ohsato, T., Takamatsu, C., Hamasaki, N. and Kang, D. (2000). The D-loop structure of human mtDNA is destabilized directly by 1-methyl-4-phenylpyridinium ion (MPP^+), a parkinsonism-causing toxin. *Eur. J. Biochem.*, *267*, 200-206.
[42] Oshige, M., Takeuchi, R., Ruike, R., Kuroda, K. and Sakaguchi, K. (2004). Subunit protein-affinity isolation of *Drosophila* DNA polymerase catalytic subunit. *Protein Expr. Purif.*, *35*, 248-256.
[43] Kusumoto, R., Masutani, C., Shimmyo, S., Iwai, S. and Hanaoka, F. (2004). DNA binding properties of human DNA polymerase η: implications for fidelity and polymerase switching of translesion synthesis. *Genes Cells*, *9*, 1139-1150.
[44] Biertümpfel, C., Zhao, Y., Kondo, Y., Ramón-Maiques, S., Gregory, M., Lee, J.Y., Masutani, C., Lehmann, A.R., Hanaoka, F. and Yang, W. (2010). Structure and mechanism of human DNA polymerase ι. *Nature*, *465*, 1044-1048.
[45] Ohashi, E., Murakumo, Y., Kanjo, N., Akagi, J., Masutani, C., Hanaoka, F. and Ohmori, H. (2004). Interaction of hREV1 with three human Y-family DNA polymerases. *Genes Cells*, *9*, 523-531.
[46] Shimazaki, N., Yoshida, K., Kobayashi, T., Toji, S., Tamai, T. and Koiwai, O. (2000). Over-expression of human DNA polymerase λ in *E. coli* and characterization of the recombinant enzyme. *Genes Cells*, *7*, 639-651.
[47] Sakaguchi, K., Hotta, Y. and Stern, H. (1980). Chromatin-associated DNA polymerase activity in meiotic cells of lily and mouse. *Cell Struct. Funct.*, *5*, 323-334.
[48] Mizushina, Y., Tanaka, N., Yagi, H., Kurosawa, T., Onoue, M., Seto, H., Horie, T., Aoyagi, N., Yamaoka, M., Matsukage, A., Yoshida, S. and Sakaguchi, K. (1996). Fatty acids selectively inhibit eukaryotic DNA polymerase activities in vitro. *Biochim. Biophys. Acta*, *1308*, 256-262.
[49] Mizushina, Y., Yoshida, S., Matsukage, A. and Sakaguchi, K. (1997). The inhibitory action of fatty acids on DNA polymerase β. *Biochim. Biophys. Acta*, *1336*, 509-521.
[50] Ogawa, A., Murate, T., Suzuki, M., Nimura, Y. and Yoshida, S. (1998). Lithocholic acid, a putative tumor promoter, inhibits mammalian DNA polymerase β. *Jpn. J. Cancer Res.*, *89*, 1154-1159.
[51] Mizushina, Y., Dairaku, I., Yanaka, N., Takeuchi, T., Ishimaru, C., Sugawara, F., Yoshida, H. and Kato, N. (2007). Inhibitory action of polyunsaturated fatty acids on IMP dehydrogenase. *Biochimie*, *89*, 581-590.
[52] Soltis, D.A. and Uhlenbeck, O.C. (1982). Isolation and characterization of two mutant forms of T4 polynucleotide kinase. *J. Biol. Chem.*, *257*, 11332-11339.
[53] Lu, B.C. and Sakaguchi, K. (1991). An endo-exonuclease from meiotic tissues of the basidiomycete *Coprinus cinereus*: Its purification and characterization. *J. Biol. Chem.*, *266*, 21060-21066.
[54] Ishiyama, M., Tominaga, H., Shiga, M., Sasamoto, K., Ohkura, Y. and Ueno, K. (1996). A combined assay of cell viability and in vitro cytotoxicity with a highly watersoluble tetrazolium salt, neutral red and crystal violet. *Biol. Pharm. Bull.*, *19*, 1518-1520.
[55] Sakaguchi, K., Sugawara, F. and Mizushina, Y. (2002). Inhibitors of eukaryotic DNA polymerases. *Seikagaku*, *74*, 244-251.

[56] Green, D.R. (2011). Means to an End: Apoptosis and Other Cell Death Mechanisms; Cold Spring Harbor Laboratory Press: New York.
[57] Patel, D., Shukla, S. and Gupta, S. (2007). Apigenin and cancer chemoprevention: progress, potential and promise. *Int. J. Oncol.*, *30*, 233-245.
[58] Hertog, M.G., Kromhout, D., Aravanis, C., Blackburn, H., Buzina, R., Fidanza, F., Giampaoli, S., Jansen, A., Menotti, A. and Nedeljkovic, S. et al. (1995). Flavonoid intake and long-term risk of coronary heart disease and cancer in the seven countries study. *Arch. Intern. Med.*, *155*, 381-386.
[59] Aherne, S.A. and O'Brien, N.M. (2002). Dietary flavonols: chemistry, food content, and metabolism. *Nutrition*, *18*, 75-81.
[60] Johannot, L. and Somerset, S.M. (2006). Age-related variations in flavonoid intake and sources in the Australian population. *Public Health Nutr.*, *9*, 1045-1054.
[61] McKay, D.L. and Blumberg, J.B. (2006). A review of the bioactivity and potential health benefits of chamomile tea (*Matricaria recutita* L.). *Phytother. Res.*, *20*, 519-530.
[62] Bevilacqua, L., Buiarelli, F., Coccioli, F. and Jasionowska, R. (2004). Identification of compounds in wine by HPLC-tandem mass spectrometry. *Ann. Chim.*, *94*, 679-689.
[63] Gerhauser, C. (2005). Beer constituents as potential cancer chemopreventive agents. *Eur. J. Cancer*, *41*, 1941-1954.
[64] Gupta, S., Afaq, F. and Mukhtar, H. (2001). Selective growth-inhibitory, cellcycle deregulatory and apoptotic response of apigenin in normal versus human prostate carcinoma cells. *Biochem. Biophys. Res. Commun.*, *287*, 914-920.
[65] Kim, H.Y., Kim, O.H. and Sung, M.K. (2003). Effects of phenol-depleted and phenol-rich diets on blood markers of oxidative stress, and urinary excretion of quercetin and kaempferol in healthy volunteers. *J. Am. Coll. Nutr.*, *22*, 217-223.

In: Apigenin and Naringenin
Editor: Nilus M. Stacks
ISBN: 978-1-63463-987-3
© 2015 Nova Science Publishers, Inc.

Chapter 6

CHEMOPREVENTION BY APIGENIN THROUGH INDUCTION OF APOPTOSIS AND AUTOPHAGY

Bokyung Sung and Nam Deuk Kim*
Department of Pharmacy, Molecular Inflammation Research Center
for Aging Intervention, Pusan National University, Republic of Korea

ABSTRACT

Apigenin (4',5,7-trihydroxyflavone), a naturally occurring flavone, is widely distributed in many fruits and vegetables, such as parsley, chamomile, celery and kumquats. In recent years, apigenin has been increasingly recognized as a cancer chemopreventive agent. The chemopreventive aspects of apigenin have been evaluated both *in vitro* and *in vivo*. Several studies in different *in vitro* and *in vivo* systems have shown that apigenin possesses the capacity to regulate response to oxidative stress and DNA damage, suppress inflammation and angiogenesis, inhibit cell proliferation, and induce autophagy and apoptosis. One of the most obvious mechanisms of apigenin is to induce p53 tumor suppressor proteins at the translational levels, followed by $p21^{WAF1/CIP1}$ induction. Another role of apigenin in chemoprevention is induction of autophagy in several human cancer cells, including HCT116 human colon cancer cells. We recently reported that apigenin suppressed growth of HCT116 cells in a concentration-dependent manner by causing G2/M cell cycle arrest. In addition, apoptosis and autophagy were induced in apigenin-treated cells. Inhibition of autophagy by 3-methyladenine enhanced the apoptosis induced by apigenin through activation of pro-caspases-8, -9, and -3 and cleavage of poly(ADP-ribose) polymerase.

These results provide evidence that inhibition of autophagy may be an effective way to improve the chemotherapy of anticancer agents against human colon cancer. Therefore, this chapter discusses the induction of apoptosis and autophagy as a mechanism of chemoprevention by apigenin.

Keywords: Apigenin, chemoprevention, apoptosis, autophagy

* Correspondence to: Nam Deuk Kim, Department of Pharmacy, College of Pharmacy, Pusan National University, 63 Beon-gil 2, Busandaehag-ro, Geumjeong-gu, Busan 609-735, Republic of Korea. Tel: +82-51-510-2801, Fax: +82-51-513-6754. Email: nadkim@pusan.ac.kr.

1. INTRODUCTION

Cancer is a public health burden, and in the United States, more than 1,665,540 new cancer cases and 585,720 cancer deaths are predicted in 2014 [1]. Growing knowledge of cancer pathobiology has led to use of a molecular approach in treatment of neoplasms and reducing cancer mortality rates; however, cancer cells develop resistance and use multiple mechanisms to escape cancer therapy. According to a recent report, the projected increase in global cancer burden—from 12.7 million new cases in 2008, to 22.2 million by 2030—indicates population growth and an evolving age distribution together with other important changes in underlying incidence, allied with the prevalence and distribution of risk factors [2]. Emerging evidence clearly proposes that primary prevention of cancer is an effective —and probably cost-effective— way to fight cancer, with between one third and one half of cancers being preventable.

Cancer chemoprevention involves the administration of one or more chemical entities, either as individual drugs or as dietary supplements, to reduce or delay the occurrence of malignancy. Chemoprevention is categorized into three kinds, primary, secondary, or tertiary. Primary chemoprevention involves the administration of agents to the general healthy population to those who may be at increased risk for disease. Secondary chemoprevention involves the identification of patients with premalignant lesions and administration of agents to prevent progression to invasive disease. Definitions of primary and secondary chemoprevention vary and some groups now combine the two scenarios under the term primary chemoprevention. Dietary phytochemicals and non-steroidal anti-inflammatory drugs (NSAIDs) are representative examples of primary chemopreventive agents. Tertiary chemoprevention is targeted for prevention of disease recurrence or second primary disease in individuals who have undergone successful treatment of early disease. At the molecular level, cancer chemoprevention is subgrouped by the disruption of, or at least the delay of, multiple pathways and processes among the three stages of carcinogenesis: initiation, promotion, and progression. Agents that inhibit the initiation stage are referred to as "blocking agents" because they "block" mutagenic interactions with DNA. Agents that affect later stages of carcinogenesis (promotion and progression) are termed "suppressing agents" for their ability to decrease the proliferative capacity of initiated cells or reduce the ability of cancer cells to evolve metastatic properties by promoting pathways leading to apoptosis. In recent years, cancer chemoprevention has emerged as a major approach for reducing cancer burden [3].

The definition of phytochemicals is chemical compounds that occur naturally in plants. It was also mentioned earlier that they are typical chemopreventive agents and exist in abundance in many fruits and vegetables. Emerging evidences indicate that the risk of cancer in general is inversely associated with the consumption of vegetables and fruits [4-6]. Numerous phytochemicals are found in fruits and vegetables, and polyphenols are the most extensively studied group of phytochemicals. Polyphenols include a large subgroup of compounds known as flavonoids, which are present in a wide range of fruits, vegetables, and grains. Several lines of evidence provide speculation that flavonoids may also be helpful in reducing cancer risk simply by functioning as antioxidants, eliminating free radicals that have been linked to DNA damage and inflammation. Previous case-control studies have shown that intake of total flavonoids, subgroups of flavonoids, or individual flavonoids is associated with a reduced risk of lung, gastric, colorectal, breast, ovarian, and endometrial cancer, and non-Hodgkin's lymphoma [7-12]. Interestingly, one case-control study evaluated the association between dietary flavonoid intake and ovarian cancer risk in 1,141 patients with ovarian cancer and 1,183 frequency-matched control subjects [10]. The individual intake of five common flavonoids (myricetin, kaempferol, quercetin, luteolin, and apigenin) and total intake of these flavonoids were calculated using questionnaires. No correlation was observed between total flavonoid consumption and cancer risk. In analyses of each individual flavonoid, only intake of apigenin showed association with a borderline significant decrease in cancer risk.

However, the results from prospective cohort studies linking flavonoid intake and cancer risk are inconsistent. Therefore, more studies are needed in order to better characterize the bioaccessibility, bioavailability, and metabolism of flavonoids; to improve our knowledge of the specific actions of each flavonoid subclass as well as individual flavonoids; and to further clarify whether flavonoids or flavonoid-rich foods may have a particularly important effect on cancer prevention.

In this chapter, we will review in detail the chemopreventive effect of apigenin. We focus on the modulation of two cellular processes for maintaining homeostasis, apoptosis and autophagy, by this powerful flavone apigenin.

2. APIGENIN

Flavonoid was discovered in the 1930s by the Hungarian biochemist Albert Szent-Györgyi. He and his coworker identified a substance from lemon peels that reduced capillary permeability and was an effective treatment in purpura patients who were resistant to vitamin C therapy [13]. Since then, scientists have discovered more than 7,000 flavonoids and the list of newly discovered flavonoids continues to grow [14].

Flavonoids, which are the largest group of polyphenols, are classified according to six classes: flavonols, flavan-3-ols, isoflavones, flavanones, flavones, and anthocyanins. With luteolin, apigenin is a represemtive compound belonging to the flavone class.

The name "apigenin", in common with many other names of flavonoids, which are related in some way to the specific epithet or genus from which the compound was originally obtained, is derived from *Apium*, a genus in Apiaceae (carrot family, also known as Umberlliferae). Because apigenin has hydroxyl groups at positions C-5 and C-7 of A-ring and C-4′ of B-ring, and belongs to a class of flavonoids known as flavones, its synonym would be 4′,5,7-trihydroxyflavone. The molecular formula of apigenin is $C_{15}H_{10}O_5$ and molecular weight is 270.24.

With flavanol kaempferol, apigenin is regarded as one of the major flavonoids because of its common occurrence in a variety of natural sources and its significant concentrations when present. Natural sources of apigenin include parsley, chamomile, celery, vine spinach, artichokes, oregano, and others (see Table 1). Among these, dried parsley is the richest source of apigenin, containing 4503.5 mg/100 g, respectively. Other sources such as chamomile (dried flower) contain 300-500 mg/100 g, celery seed 78.65 mg/100 g, vine spinach 62.2 mg/100 g, and Chinese celery 24.02 mg/100 g. Other than apigenin, various glycosylated derivatives (e.g., apiin and apigetrin) and dimers of apigenin such as amentoflavone (3′,8″-biapigenin) have been found in natural sources.

Table 1. Selected dietary sources, shape and parts with the highest amount of apigenin

Common name	Description	Scientific name	Content (mg/100 g)
Parsley	Dried	*Petroselinum crispum*	4503.5
	Fresh		215.46
Chamomile	Dried flower	*Matricaria chamomilla*	300~500
Celery	Seed	*Apium graveolens*	78.65
	Hearts, green		19.1
Vine spinach	Raw	*Basella alba*	62.2
Chinese celery	Raw	*Apium graveolens* var. *secalinum*	24.02
Kumquats	Fruits	*Fortunella* spp.	21.87
Mexican oregano	Dried	*Lippia graveolens*	17.71
Queen Anne's lace	Leaves	*Daucus carota*	12.6
Artichoke	Fresh	*Cynara scolymus*	7.48
Juniper berries	Green, unripe	*Juniperus communis*	7.26
Peppermint	Fresh	*Mentha x piperita* L. *nothosubsp. Piperita*	5.39
Rutabagas	Raw	*Brassica napus* var. *napobrassica*	3.85
Annual saw-thistle	Leaves	*Sonchus oleraceus*	3.8
Marjoram	Dried	*Origanum majorana*	3.5
Oregano	Fresh	*Origanum vulgare*	2.57
Sorghum	Grain	*Sorghum bicolor*	2.54
Thyme	Fresh	*Thymus vulgaris*	2.5
Celeriac	Raw	*Apium graveolens* var. *rapaceum*	2.41
Peppers (hot chili)	Green	*Capsicum frutescens*	1.4

From "USDA Database for the Flavonoid Content of Selected Foods, Release 3.1"; http://www.usda.gov/wps/portal/usda/usdahome

3. Anticancer and Chemopreventive Effects of Apigenin

In recent decades, plants with high content of apigenin such as chamomile or parsley have been used in treatment of a variety of disorders, including inflammatory diseases [15, 16]. In the 1960s, apigenin first gained scientific interest, due to its anti-allergic effect [17]. In the 1980s, Birt and colleagues [18] demonstrated anti-mutagenic and anti-promotion properties of apigenin. They found that apigenin effectively blocked both initiation and promotion of carcinogenesis [18]. Since then, extensive research over the past several years has indicated the anti-cancer and chemopreventive effect of apigenin in various animal models of cancer (Table 2).

Table 2. *In vivo* anticancer and chemopreventive effects of apigenin

Models	Effects and mechanisms	Dose and route of administration	Refs.
Breast			
BT-474 xenograft	Inhibits progression and development of tumors by inducing apoptosis, inhibiting cell proliferation and reducing Her2/neu.	50 mg/kg/day, i.p. for 20 days	[19]
MPA-accelerated DMBA-induced mammary carcinogenesis (rat)	Delays the development of, and decreases the incidence and multiplicity of mammary tumors; decreases VEGF and suppresses VEGFR2 expression.	50 mg/kg/day, i.p. for 10-13 days	[20]
Colon			
HCT116 xenograft	Inhibits tumor growth; decreases Mcl-1, p-Akt and p-ERK.	25 mg/kg/day, gavage for 3 weeks	[21]
SW480 orthotopic	Inhibits growth and metastasis of CRC.	50 mg/kg/day for 6 weeks	[22]
SW480 xenograft	Inhibits tumor growth; promotes apoptosis through FADD.	20 mg/kg/day, i.p. for 27 days	[23]
DHDK 12 syngenic	Decreases tumor volume.	50 mg/kg/day, i.p. for 9 days	[24]
AOM-induced carcinogenesis (rat)	Lowers the number of high multiplicity ACF; increases apoptosis of luminal surface colonocytes.	0.1% in diet for 10 weeks	[25]
AOM/bombesin-induced carcinogenesis (rat)	Decreases the incidence of cancer metastasis and lymphatic vessel invasion of adenocarcinomas.	0.75, 1.5 mg/kg, s.c., every other day for 29 weeks	[26]
$APC^{Min/+}$ mice	Reduces polyp numbers accompanied by increasing p53activation through phosphorylation.	25, 50 mg/kg, gavage, 14 times every other day for 1 month	[27]
Glioma			
C6 xenograft	Decreases tumor volume.	50 mg/kg/day injection for 12 days	[24]
Leukemia			
U937 xenograft	U937 xenograft	U937 xenograft	U937 xenograft

Table 2. (Continued)

Models	Effects and mechanisms	Dose and route of administration	Refs.
Lung			
A549 xenograft*	Inhibits tumor growth and expression of HIF-1α, VEGF and PCNA; induces apoptosis in tumor.	3 mg/kg, i.p., three times a week for 36 days	[29]
LLC syngenic	Decreases tumor volume slightly.	50 mg/kg/day injection for 12 days	[24]
Liver			
Huh7 xenograft	Impairs tumor growth.	50 μg/mouse/day, i.p. for 4 weeks	[30]
NDEA-induced carcinogenesis (rat)	Decreases tumor incidence and tumor nodules; decreases lipid peroxidation, protein carbonylation and DNA damage.	25 mg/kg/day, i.p. for 14 days	[31]
	Increases effect of cisplatin on tumor growth.	25 mg/kg/day, i.p. for 19 days	
Oral			
DMBA-induced oral carcinogenesis (hamster)	Lowered tumor incidence, invasion and tumor volume; decreases p53 and PCNA.	1.1 mg/15 ml in drinking water for 12 weeks	[32]
	Reduces tumor incidence.	0.16 mg/day, s.c. for 11 weeks	[33]
	Prevents the tumor incidence, volume and tumor burden; downregulates apoptotic, cell proliferative and angiogenic markers; modulates antioxidants and phase I and phase II detoxification cascades.	2.5 mg/kg/day, oral for 15 weeks	[34, 35]
Osteosarcoma			
U-2 OS xenograft	Reduces tumor mass and decreased the tumor weight.	2 mg/kg, i.p., once every 3 days for 30 days	[36]
Ovary			
A2780 orthotopic	Inhibits tumor growth and metastasis; decreased FAK protein level of the primary tumors.	5 mg/kg/day, i.p. for 4 weeks	[37, 38]
Pancreas			
MiaPaCa-2 xenograft	Enhances therapeutic effect of gemcitabine; inhibits tumor growth; induces apoptosis and blocks Akt-NF-κB pathway.	50 mg/kg, i.p., five times/week for 3 weeks	[39]
MiaPaCa-2 orthotopic	Reduces tumor volume; increases apoptosis in tumor tissue.	0.2% apigenin-supplemented diet for 6 weeks	[40]

Models	Effects and mechanisms	Dose and route of administration	Refs.
Prostate			
PC-3 xenograft	Decreases tumor volume and wet weight; induces apoptosis through reduction of XIAP and survivin, induction of p27, p21, Bax/Bcl-2 ratio, and cleavage of Bax, caspase-9 and PARP; inhibits HDAC1 activity and expression; decreases the expression of IGF-IR,p-IGF-IR, p-Akt, p-GSK-3β, p-BAD, cyclin D1 and PCNA; modulates cell cycle regulatory proteins.	20, 50 μg/mouse/day, gavage for 8 weeks	[41-45]
22Rv1 xenograft	Impairs tumor growth and reduced wet weight; triggers apoptosis through induction of p53, p-p53, cytochrome *c*, Bax and cleaved caspase-3; decreases Bcl-2; induces IGFBP-3; reduces IGF-I levels in serum; induces IGFBP-3 mRNA expression; modulates cell cycle regulatory proteins.	20, 50 μg/mouse/day, gavage for 10 weeks	[45-47]
TRAMP mice	Decreases tumor volumes and abolishes distant organ metastasis; reduces p-Akt and p-FoxO3a, expression of Ki-67 and cyclin D1; upregulates BIM and p27.	20, 50 μg/mouse/day, gavage for 20 weeks	[48]
Skin			
B16-BL6 syngenic	Delays tumor growth; inhibits lung colonization.	25, 50 mg/kg/day, i.p. for 19 days	[49]
UVB-induced skin inflammation	Decreases UVB-induced ear edema and COX-2 expression in mouse skin.	7.5, 30 nmol in acetone (0.2 ml), topical, once every 2 days for total of four times	[50]
DMBA-TPA-induced skin tumorigenesis	Reduces incidence and number of papillomas.	20 μmol (0.2 ml), topical, twice a week for 20 weeks	[51]
	Reduces incidence and number of carcinomas.	5, 20 μmol (0.2 ml), topical, twice a week for 20 weeks	

*, For the apigenin group, the cells were injected into the mice with 15 μM apigenin

ACF, aberrant crypt foci; AOM, azoxymethane; BAD, Bcl-2-associated death promoter; Bax, Bcl-2-associated X protein; BIM, Bcl-2-interacting mediator of cell death; COX-2, cyclooxygenase-2; CRC, colorectal cancer; DEN, diethylnitrosamine; DMBA, 7,12-dimethylbenz(*a*)anthracene; ERK, extracellular signal-regulated kinases; FADD, Fas-associated protein with death domain; FAK, focal adhesion kinase; FoxO3a, Forkhead box O3a; GSK-3β, glycogen synthase kinase 3 beta; HDAC1, histone deacetylase 1; HIF-1α, hypoxia inducible factor 1 alpha; i.p., intraperitoneal; IGF-1, insulin-like growth factor-1; IGFBP, IGF binding protein; IGF-IR, type I IGF receptor; JNK, c-Jun N-terminal kinases; Mcl-1, myeloid cell leukemia-1; MPA, medroxyprogesterone acetate; NDEA, N-nitrosodiethylamine; NF-κB, nuclear factor kappa-light-chain-enhancer of activated B cells; PARP, poly(ADP-ribose) polymerase; PCNA, proliferating cell nuclear antigen; s.c., subcutaneously; TPA, 12-*O*-tetradecanoylphorbol-13-acetate; TRAMP, transgenic adenocarcinoma of the mouse prostate; UVB, ultraviolet B; VEGF, vascular endothelial growth factor; XIAP, X-linked inhibitor of apoptosis protein.

Briefly, treatment with apigenin (20 and 50 μg/mice) for 8 weeks resulted in decreased tumor volume and wet weight of tumors of prostate cancer cells in nude mice without any significant effect on body weight [41, 42, 47] and produced chemopreventive effects in the transgenic adenocarcinoma of a mouse prostate (TRAMP) mouse model [48]. Apigenin was found to cause apoptosis and elevate active caspase-3 in tumor cells from xenograft [41, 42, 47]. Apigenin exhibited significant antitumor activity in a human colon cell implanted nude mice model [21-23]. Anti-tumorigenic activity of apigenin has also been reported in an orthotopic colorectal nude mouse model [22]. In addition, apigenin significantly inhibited tumor growth, as well as distant organ metastasis. This flavone (0.1% in diet) also increased apoptosis of luminal surface colonocytes and reduced the incidence of aberrant crypt foci, one of the earliest precursors of colorectal adenoma development, particularly the tumor initiation phase [25]. Similarly, oral administration of apigenin reduced polyp numbers, accompanied by activation of p53, a tumor suppressor gene, in $APC^{Min/+}$ mice [27]. Apigenin was also found to induce apoptosis, inhibit cell proliferation, and reduce Her2/neu in breast cancer BT-474 xenografts in a mouse model [19]. Likewise, intake of an apigenin containing diet (0.2%) for 6 weeks produced significant antitumor activity by inducing apoptosis in an orthotopic pancreatic nude mouse model [40]. Oral administration of apigenin (2.5 mg/kg) for 15 weeks in 7, 12-dimethyl-ben z[*a*]anthracene-induced oral carcinogenesis in hamsters led to reduced volume and tumor incidence, downregulation of protein markers associated with apoptosis, cell proliferation and angiogenesis, and modulation of phase I and II detoxification cascades, thereby suggesting that apigenin can act as an excellent chemopreventive agent in overcoming oxidative stress-mediated diseases like cancer [34, 35]. Wei et al. [51], who investigated the potential chemopreventive effects of apigenin in murine skin carcinogenesis, found that it caused marked reduction of incidence and number of papillomas as well as carcinomas. Byun and colleagues [50], who also explored beneficial effect of apigenin in ultraviolet B (UVB)-induced skin inflammation, found that flavones reduced ear edema and inflammatory mediator cyclooxygenase-2 expression in mice. Overall, a vast majority of these studies indicate that apigenin can inhibit tumor initiation, progression, and metastasis in a wide variety of preclinical cancer models.

3.1. Role of Apigenin in Apoptosis

3.1.1. Apoptosis

Programmed cell death, also known as apoptosis, is a tightly tied fundamental cellular program that plays critical roles in development and homeostasis of multicellular organisms. Because tissue homeostasis is maintained by a balance between cell death on one side and cell proliferation on the other side, disruption of this balance, for example, dysregulated apoptosis, has been implicated in a variety of human diseases including cancer [52, 53]. Apoptosis is triggered through two principal pathways. One is the extrinsic pathway, depending on triggering of death receptors on the cell surface, and the other is the intrinsic pathway, mediated by release of mitochondrial intermembrane space proteins into cytosol. The extrinsic apoptosis pathway is initiated through the ligation of death receptors by their cognate ligands. Death receptors (DRs) comprise CD95 (Fas/Apo1), tumor necrosis factor (TNF) receptor as well as TNF-related-apoptosis-inducing ligand (TRAIL) receptors and their corresponding ligands are CD95 ligand, TRAIL, and TNF-α. Upon engaging with their ligands, DRs formed death-inducing signaling complex (DISC) by recruitment of adaptor molecules leading to activation of caspase-8. The activated caspase-8 can either directly activate effector caspases such as caspase-3 or indirectly initiate activation of the intrinsic pathway of apoptosis [54]. Within the intrinsic apoptosis, physical or chemical apoptotic stimuli triggered release of mitochondrial proteins (e.g., cytochrome *c*, second mitochondria-derived activator of caspases) into cytosol. The released cytochrome *c* activates caspases by forming apoptosome, a protein complex composed of cytochrome *c* and apoptotic peptidase activating factor 1 (Apaf-1), leading to caspase-9 and subsequently caspase-3 activation [55].

Based on the crucial role of apoptosis, rational targeting of apoptosis has important implications cancer therapy, and if any agents, for example apigenin here, can modulate this pathway, it will provide a new avenue for cancer treatment. The various molecular targets modulated by apigenin are summarized in Table 3.

Table 3. Molecular targets of apigenin-induced apoptosis

Cancer/cell lines	Up-regulation	Down-regulation	Refs.
Bladder			
T24	PARP & caspase-3 cleavage, Bax, Bad	p-Akt, PDK, PI3K, Bcl-2, Bcl-xL	[56]
Breast			
SK-BR-3	p53, p21, Bax, Cyt *c*	cyclin A, B, D, & E, CDK1	[57]
MDA-MB-231	p-p53 (Ser-15), p21, Bax, caspase-3 & PARP cleavage	Bcl-xL, cyclin B1, Bcl-2	[58, 59]
MCF-7 (*HER2* overexpressing)	p53, p-p53 (Ser-15), p21, caspase-8 & PARP cleavage	p-MDM2, p-JAK1, p-STAT3, NF-κB/p65, p-IκBα	[60]
MDA-MB-453	caspase-3, -6, -7, -8 & PARP cleavage, Cyt *c* release	procaspase-9, p-JAK2, p-STAT3	[61, 62]
T47D	caspase-3 & PARP cleavage, Bax	Bcl-2, Bcl-xL	[59]
Cervical			
HeLa	p53, p21, caspase-3 activity, Fas	Bcl-2	[63]
Colon			
HCT-116	p21, p53, NAG-1, Bim-EL, Bim-L, PARP cleavage	cyclin B1, Cdc2, Cdc25c, procaspase-3, -8, -9, Mcl-1, p-Akt, p-ERK	[21, 27, 64]
LoVo	p21, NAG-1		[27]
DLD-1		Mcl-1, p-Akt, p-ERK	[21]
SW480		Cdc2, cyclin B1	[65]
HT-29 & Caco2		Cdc2	[65]
Esophageal			
KYSE-510	p21, PIG3, p63, p73	cyclin B1	[66]
Gastric			
HGC-27 & SGC-7901	Bax, Bcl-2, caspase-3	MMP	[67]
Head & Neck			
SCC25	TRAIL, TRAIL-R1 & -R2, Fas, TNF-α, TNF-R1 & -R2, Bax, caspase-3	Bcl-2	[68]

Table 3. (Continued)

Cancer/cell lines	Up-regulation	Down-regulation	Refs.
Melanoma			
A375	Cyt *c* release, Bax, Apaf-1, caspase-3, -9 & PARP cleavage	Bcl-2	[69, 70]
Leukemia			
THP-1	caspase-3 activity, p-p38, p-ERK, PKCδ activity, p-ATM	caspase-9 activity, p-H2AX	[71, 72]
U937	caspase-3, -7, -9 & PARP cleavage, p-JNK, Bcl-2 cleavage	hTERT, c-Myc, Mcl-1, p-Akt, p-Bad, p-mTOR, p-GSK3β	[28, 73]
HL60	p-Cdc2, p-p38, caspase-3, -8 & PARP cleavage	PI3Kp85, p-Akt, p-GSK3β, p-JAK2, p-Src, p-STAT3	[74, 75]
TF1	LMWPTP	CDK6, p-Src, p-JAK2, p-SHP2, p-STAT3 & 5, p-p70S6K	[74]
Liver			
Huh-7	caspase-3,-8 & -9 cleavage		[76]
Lung			
A549	p21, Cyt *c* release, Bax, p53, wee1, chk2, Bid, GRP78, caspase-3, -9 & PARP cleavage, GADD153, AIF	XIAP, Bcl-2, MMP, cyclin B, cdc25c, procaspase-8	[69, 77, 78]
H460	p21, Bax, FasL, p53, AIF, Cyt *c*, caspase-3, GRP78, GADD153	XIAP, Bcl-2, Bid, procaspase-8	[77, 79, 80]
Cancer/cell lines	Up-regulation	Down-regulation	Refs.
Lymphoma			
PEL	caspase-3, -9 & PARP cleavage, Cyt *c* release	Bad-14-3-3 interaction, Bcl-2, MMP, cIAP-1, survivin, XIAP	[81]
Multiple_myeloma			
U266 & RPMI 8226	PARP cleavage	Mcl-1, Bcl-2, Bcl-xL, XIAP, survivin	[82]
Neuroblastoma			
SK-N-DZ & SK-N-BE2	caspase-3, -8 & PARP cleavage, Bax, Bid, tBid	N-Myc, E-cadherin, Notch-1, hTERT, PCNA	[83]
NUB-7	PARP cleavage, p53 (NE), p21, Bax, p-ERK		[84]
Oral			
SCC-25		cyclin D1& E, CDK1	[85]
Osteosarcoma			
U-2 OS	Bax, PARP cleavage, p53, AIF	procaspase-3, -8, & -9, GADD153 (NE)	[36]

Cancer/cell lines	Up-regulation	Down-regulation	Refs.
Pancreatic			
BxPC-3	Ac-p53, p21, PUMA, Cyt *c* release, caspase-3 cleavage	Bcl-xL/p53 interaction, Bcl-xL/PUMA interaction, cyclin B1, Bcl-2, XIAP, p-GSK3β, NF-κB/p65 (NE)	[40, 86]
MiaPaCa-2	Ac-p53, p21, PUMA, Cyt *c* release, PARP cleavage	Bcl-xL/p53 interaction, Bcl-xL/PUMA interaction	[40, 87]
PANC-1	Cyt *c* release, caspase-3 cleavage	cyclin B1, XIAP, p-GSK3β, NF-κB/p65 (NE)	[86]
Prostate			
22Rv1	p53, p-p53, p21, p14, Cyt *c* release, Bax, Apaf-1, caspase-3, -8, -9 & PARP cleavage	MDM2, MMP, Bcl-2, Bcl-xL	[46]
PC-3	caspase-3,-9 & PARP cleavage, Bax, Bad, Ku70, Cyt *c* release, p27	XIAP, cIAP-1 &-2, Bcl-2, Bcl-xL, survivin, HDAC1, procaspase-3, -7 & -9, cyclin D1	[41, 43, 88]
LNCaP	p21, p27, Bax, PARP cleavage, Cyt *c* release	cyclin D1, D2 & E, CDK2, 4 & 6, Bcl-2, procaspase-3, -8 & -9, NF-κB/p65	[88, 89]
DU145	caspase-3, -9, & PARP cleavage, DR5, Cyt *c* release	XIAP, cIAP-1 & -2, survivin, procaspase-3, -7 & -9	[41, 88, 90]
Thyroid			
FRO	c-Myc, Bid, Fas, p-p53, caspase-3 & PARP cleavage	Bcl-2, p27, p21	[91]

AIF, apoptosis-inducing factor; Apaf-1, apoptotic protease activating factor-1; ATM, ataxia telangiectasia mutated; Bad, Bcl-2-associated death promoter; Bax, Bcl-2 associated X protein; Bcl-2, B-cell lymphoma-2; Bcl-xL, B-cell lymphoma extra-large; Bid, BH3-interacting-domain death agonist; Bim-EL, Bcl-2-interacting mediator of cell death (Bim)-extra long; Bim-L, Bim-long; Cdc2, cell division control protein 2; Cdc25c, cell division cycle 25c; CDK, cyclin-dependent kinase; Chk2, checkpoint kinase 2; cIAP, cellular inhibitor of apoptosis protein; Cyt *c*, cytochrome *c*; ERK, extracellular signal-regulated protein kinases; FasL, apoptosis stimulating fragment (Fas) ligand; GADD153, growth arrest- and DNA damage-inducible gene 153; GRP78, glucose-regulated protein 78; GSK-3β, glycogen synthase kinase-3 beta; H2AX, histone H2A, X; HDAC, histone deacetylase; hTERT, human telomerase reverse transcriptase; JAK, janus family of tyrosine kinase; JNK, c-Jun N-terminal kinases; LMWPTP, low-molecular-weight protein tyrosine phosphatase; MAPK, mitogen-activated protein kinase; Mcl-1, myeloid cell leukemia-1; MDM2, mouse double minute 2; MMP, mitochondrial membrane potential; NAG-1, nonsteroidal anti-inflammatory drug (NSAID)-activated gene-1; NE, nuclear extract; p70S6K, 70-kDa ribosomal protein S6 kinase; PARP, poly(ADP-ribose) polymerase; PCNA, proliferating cell nuclear antigen; PDK-1, phosphoinositide-dependent protein kinase 1; PEL, primary effusion lymphoma; PI3K, phosphoinositide 3-kinase; PIG3, p53 induced gene 3; PKB, protein kinase B; PKC, protein kinase C; PTEN, phosphatase and tensin homolog; PUMA, p53-upregulated modulator of apoptosis; S6, S6 ribosomal protein; SHP-2, Src homology 2 domain-containing phosphotyrosine phosphatase 2; STAT, signal transducer and activator of transcription; TNFR, TNF receptor; TNF-α, tumor necrosis factor alpha; TRAIL, TNF-related apoptosis-inducing ligand; TRAIL-R, TRAIL receptor; XIAP, X-linked inhibitor of apoptosis protein.

3.1.2. Caspase-Mediated Apoptosis

The caspases, a family of cysteine proteases, provide crucial links in cellular networks controlling inflammation and cell death. To date, more than a dozen caspases have been discovered; caspase-3, -6, -7, -8, and -9 are known to be involved in apoptosis. These enzymes are again subclassified according to their mechanism of action and are either initiator caspases (caspase-8 and -9) or effector (or executioner) caspases (caspase-3, -6, and -

7) [92]. Initiator caspases activate effector caspases that subsequently coordinate their activities to demolish key structural proteins and activate other enzymes. Caspase activation is mediated by either the intrinsic or extrinsic pathway and these pathways can also crosstalk through caspase-8. Therefore, function and expression of caspase are downregulated in tumors, suggesting that activation of caspase may be an efficient strategy in cancer therapy [93].

Activation of caspases and caspase-dependent apoptotic cell death by apigenin have been substantiated in a wide variety of tumor cell lines including head and neck cancer (FaDu) [94], thyroid carcinoma (FRO) [91], gastric cancer (HGC-27, SGC-7901) [67], cervical (HeLa) [63], neuroblastoma (SK-N-DZ, SK-N-BE2) [83, 84], colon (HCT116) [64], leukemia (THP-1, U937, HL60) [28, 71-74], pancreatic cancer (MiaPaCa-2) [87] and bladder cancer (T24) [56] and lung cancer (H460, A549) [78, 80].

Numerous studies have demonstrated the apoptotic effect of apigenin through caspase activation in breast cancer cells. Apigenin activated caspase-8, caspase-9, and caspase-3, and induced the cleavage of poly(ADP-ribose) polymerase (PARP), which in turn resulted in apoptosis in HER2-overexpressing MDA-MB-453 breast cancer cells [61, 62, 95]. The apoptotic cell death mediated through apigenin-induced caspase-3 activation was also found in T47D and MDA-MB-231 breast cancer cells [59, 96]. In addition, apigenin triggered apoptotic cell death in caspase-3 deficient MCF-7 cells; this might be explained by activation of caspase-8 by apigenin, resulting in proteolytic cleavage of PARP [60].

The plant flavone apigenin was shown to induce apoptosis in a caspase-mediated manner in human prostate cancer cells. Apigenin was found to induce apoptosis in a dose-dependent manner in PC3 and DU145 cells through caspase activation, as confirmed by the use of caspase-9-specific inhibitor, zLEHD-fmk, and a general caspase inhibitor, zVAD-fmk [44]. Elevations of cleaved caspase-9 and -3 in tumor tissues from apigenin-administrated mice were also observed in the same study [44]. A similar effect was also observed in 22Rv1 human prostate carcinoma epithelial cells treated with apigenin [46]. In another study, Das et al. [69] showed that reduction of cytochrome *c* levels in mitochondrial fraction following increase in caspase-3 and -9 by apigenin is responsible for induction of apoptotic cell death in A375 melanoma cells. In addition, apigenin with poly(lactide-co-glycolide) (PLGA)-loaded nanoparticle (NAp) was shown to produce an enhanced effect in inducing cell death and modulation of cytochrome *c* release, expression of Apaf-1, Bax, Bcl-2, caspase-9, -3, and PARP cleavage in A375 cells [70]. The same group also demonstrated that NAp exerted better anti-carcinogenic effects than free apigenin in UVB and benzo(*a*)pyrene (BaP)-induced mice skin tumor model [97]. Dietary or topical administration of NAp resulted in reduced tumor incidence with no significant difference in tumor multiplicity; however, combined oral and topical treatment with NAp resulted in reduction of skin tumor multiplicity. Apigenin-nanoparticle induced cell death or apoptosis mediated by activation of caspase-3 in tumor cells thereby contributes to the inhibition of UVB-induced skin tumor growth.

3.1.3. c-myc Oncogene-Mediated Apoptosis

The proto-oncogene *c-myc* was discovered in the late 1970s as the cellular homolog of the retroviral *v-myc* oncogene [98]. It has a pivotal function in proliferation, transformation, and differentiation of cells and regulates gene transcription. In addition, c-Myc, the product of the *c-myc* proto-oncogene, also plays a role in sensitization of cells to various proapoptotic stimuli such as genotoxic stress, hypoxia, nutrient deprivation, and death receptor signaling

[99]. Studies on c-Myc have demonstrated that the role of c-Myc in apoptosis is double-edged. In the early 1990s, it was observed that ectopic expression of *c-myc* promoted apoptosis in cells deprived of survival factors. However, other observations in B cells have suggested that c-Myc may regulate apoptosis in the opposite fashion: inhibition of c-Myc resulted in dramatic apoptosis, whereas overexpression of *c-myc* protected B cells from apoptosis. Although the precise mechanisms of how c-Myc is involved in apoptosis remain unclear, it appears that multiple pathways are regulated by c-Myc. Deregulated *c-myc* upregulates p14/ARF (alternate reading frame), which in turn activates p53 to regulate a group of target genes that activate apoptosis and cell cycle arrest [99].

Apigenin has been reported to augment c-Myc in cancer cells [91]. In FRO anaplastic thyroid carcinoma (ATC) cells, apigenin-induced c-Myc acts as a core regulator in conjunction with phosphorylation of p38 and p53 and is necessary for apigenin-induced apoptosis in FRO ATC cells [91]. Apigenin also inhibits epidermal growth factor receptor (EGFR) tyrosine autophosphorylation and phosphorylation of its downstream effector mitogen activated protein kinase (MAPK), which is associated with cell proliferation of ARO human anaplastic thyroid cancer cells [100]. The inhibitory effect of apigenin on the EGFR pathway caused a reduction of phosphorylated c-Myc, a nuclear substrate for MAPK, finally leading to programmed cell death involving DNA fragmentation. A study conducted by Shukla et al. [101] showed that p.o. administration of apigenin reduced nuclear level of c-Myc in DU145 prostate cancer cells and the dorsolateral prostates of TRAMP mice.

Unlike thyroid and prostate cancer cells, apigenin triggered apoptosis through c-Myc downregulation in U937 leukemia cells [73]. Inhibition of c-Myc expression as well as its DNA-binding activity on human telomerase reverse transcriptase (hTERT) promoter site by apigenin resulted in a reduction of hTERT expression lead to the loss of telomerase activity.

3.1.4. Tumor Suppressor p53-Dependent Apoptosis

The mechanisms involved in apigenin-induced apoptosis appear to be multiple, including both p53-dependent and -independent biochemical and molecular cascades. The most thoroughly studied tumor suppressor gene, p53, described as the "guardian of the genome", is regarded as a key element in maintaining balance between cell growth and cell death in the living system [102, 103]. This protein has been known to be involved in cellular responses to a wide range of cellular stress, including DNA damage, leading to triggering of cell cycle regulatory events to limit the proliferation of damaged cells [104, 105]. The ability of p53 to control apoptosis is one of the most extensively studied areas, and these studies indicate that apoptosis contributes to tumor suppressor activity of p53. As a proapoptotic mediator, p53 can activate transcription of various proapoptotic genes, including those encoding members of the Bcl-2 family such as the BH-3 only proteins Bax, Noxa, and Puma. Alternatively, p53 can also induce apoptosis by suppression of antiapoptotic genes such as *survivin*, thus promoting caspase activation [106]. Other apoptosis inducing gene products, including Fas, TRAIL receptor DR5, Bid, and Apaf-1, have also been known to be regulated by p53 [107, 108]. Collectively, the above mentioned functions of p53 provide the rationale that p53 modulating reagent may represent a promising therapy for cancer treatment.

Several studies have indicated that apigenin increases the accumulation of p53, further enhances p53-dependent $p21^{WAF1/CIP1}$, which caused cell cycle arrest and apoptosis in various cancer types [57, 63, 64, 89, 109]. Torkin and colleagues [84] reported that apigenin caused apoptosis in human neuroblastoma cells, not in nontransformed cells. The mechanism of action of apigenin seems to involve p53, as it increased the levels of p53 and the p53 target genes, $p21^{WAF1/CIP1}$ and Bax. In addition, apoptotic cell death mediated by apigenin occurred in wild-type p53 harboring cells but not in cells with nonfunctional mutant p53 suggesting that apigenin requires wild-type p53 for induction of apoptosis. Shukla et al. [46] demonstrated that the p53-associated pathway is required for apigenin-mediated apoptosis, as evidenced by the p53 antisense oligonucleotide experiment. Treatment of prostate cancer 22Rv1 cells with apigenin could increase the expression and transcriptional activation of p53, thus this increased p53 protein expression showed correlation with an increase in the levels of its transcriptional target $p21^{WAF1/CIP1}$. Further, consistent with the findings *in vitro*, apigenin intake resulted in a dose-dependent increase in expression of wild-type p53, p53-Ser15 phosphorylation, cytochrome *c*, and cleaved caspase-3 in 22Rv1 transplanted nude mice. An increase of Bax and a decrease in Bcl-2 levels were observed after apigenin administration, indicating that inhibition of growth of 22Rv1 tumor xenografts is due to induction of apoptosis, which is mediated *via* p53 pathways. In addition to apigenin, one of its glucosides, vitexin (apigenin-8-C-D-glucopyranoside), has been shown to suppress metastasis and induce apoptosis through p53-dependent pathway in human oral cancer OC2 cells [110].

Activity of p53 is largely regulated by posttranslational modifications [111]. Phosphorylation of p53 has been shown to play important roles in regulation of biological activities of p53. Indeed, phosphorylation of p53 at Ser-15 has been shown to be critical for the apoptotic activity of p53 and not for p53 stabilization [112]. In studies by Kim et al. [91] treatment with apigenin resulted in multiplied phospho-p53 protein levels, but decreased $p27^{KIP1}$ and $p21^{WAF1/CIP1}$ protein levels in FRO thyroid carcinoma cells. A study conducted by Zhong et al. [27] reported association of apoptosis induced by apigenin with induction of pro-apoptotic protein p53, $p21^{WAF1/CIP1}$ and NSAID-activated gene-1 (NAG-1), and an increase of p53 phosphorylation at Ser-15 in HCT116 cells. In agreement with the *in vitro* results, apigenin reduced polyp numbers, accompanied by increasing p53 activation through phosphorylation in animal models of tumor tissues from $APC^{Min/+}$ mice. Cai and Liu [113] reported that apigenin specifically inhibited Thr-55 phosphorylation of p53, which is required for p53 nuclear export, blocked the chromosome region maintenance 1 (CRM1)-p53 association, restored p53 nuclear localization, and sensitized neuroblastoma cells with cytoplasm localized wild-type p53 to DNA damage.

3.1.5. p53-Independent Apoptosis

In addition to its role as a tumor suppressor, p53-dependent apoptosis contributes to chemotherapy-induced cell death. However, the loss of p53 function was linked to chemoresistance in certain tumor types, and cancers with a p53 mutation generally respond poorly to therapeutics. With regard to human cancer, mutation of p53 is frequently found in at least 50% of all human cancers and results in uncontrolled cellular proliferation [114]. In this context, anti-cancer agents working independent of p53 status have also been studied. Emerging evidence has shown that several apoptosis inducing agents exert their activity in a p53-independent manner. The dependence on p53 seems to be determined by the status of p53 in cell types (i.e., p53-deficiency, p53-mutation) and the apoptotic signal triggered [99]. The flavone apigenin induced apoptosis through the mitochondrial death pathway and induction of $p21^{WAF1/CIP1}$ in p53 mutation harboring human esophageal squamous cell carcinoma KYSE-510 cells [66]. Another study explored the cytotoxic effect of apigenin against OE33 human esophageal adenocarcinoma cells. Apigenin induced p53-independent mitochondrial-mediated apoptosis through up-regulation of p53 induced gene 3 and cleavage of caspase-9 and caspase-3 [115]. Treatment of prostate cancer DU145 (with mutated p53) and PC-3 (with null p53) cells with apigenin resulted in an increase in protein expression of $p21^{WAF1/CIP1}$ and induction of apoptosis. These results suggest that apigenin exerts a chemopreventive effect on cancer cells in a p53-independent fashion [45, 116, 117]. In human pancreatic cancer cells, there was little reversal of the anti-proliferative and pro-apoptotic effects of apigenin when transcription-dependent p53 activation was blocked with pifithrin-α, a specific inhibitor of p53 DNA binding, while upregulation of p53-regulated apoptotic genes $p21^{WAF1/CIP1}$ and Puma was suppressed by pifithrin-α. These data indicate that apigenin may activate p53 through a parallel, transcriptionally independent pathway of programmed cell death [40]. A study conducted with malignant breast cancer MDA-MB-231 cells (with high level of mutant p53) showed that apigenin induced apoptosis in a p53-independent manner. Apigenin increased the level of active p53 (phosphorylation at Ser 15) without changing the p53 level and upregulated $p21^{WAF1/CIP1}$, indicating that apigenin abrogated the tumor-promoting effect of mutant p53 and introduced a p53-independent regulatory mechanism to induce growth suppression and apoptosis in MDA-MB-231 cells [58].

3.1.6. Proteasome-Mediated Apoptosis

The ubiquitin–proteasome system (UPS) maintains homeostasis and integrity of cellular processes by regulating protein degradation. UPS has a pivotal role in cell cycle control, transcriptional regulation, signal transduction, inflammation, endoplasmic reticulum-mediated degradation, membrane trafficking, receptor endocytosis, apoptosis, and development [118]. Ubiquitin (Ub) is attached to lysine residues on the protein substrate via multiple ATP-dependent processes by Ub-activating (E1), Ub-conjugating (E2), and Ub-ligating (E3) enzymes. These enzymes are harmoniously coordinated to link ubiquitin moiety to specific targets and regulate oligomerization, degradation, and posttranslational modification of target proteins. Proteasome is composed of a multicatalytic enzyme complex containing one catalytic core, the 20S proteasome, and two 19S regulatory subunits. The proteolytic activity of the proteasome resides in the 20S proteasomal subunits, β1, β2, and β5, which are responsible for caspase-, trypsin-, and chymotrypsin-like activities, respectively [119]. Proteasome activity is often elevated in cancers, consistent with the requirement for this proteolytic system in rapidly growing tissues [120]. Substantial evidence indicates that inhibition of proteasome activity resulted in accumulation of several proteins (i.e., IκBα, Bax, cyclins, p53, and $p27^{KIP1}$) involved in carcinogenesis and cancer survival [121, 122]. In this regard, proteasome inhibition has become increasingly important in anticancer research for these reasons.

Apigenin possesses inhibitory effects against proteasome, with its potency being chymotrypsin-like activity [96, 123]. Inhibition of the proteasome activity by apigenin was associated with breast cancer cell apoptosis *in vitro* and regression of tumor growth in nude mice [96]. This flavonoid was reported to induce increased accumulation of ubiquitinated proteins such as Bax and IκBα by inhibition of proteasome activity, thereby inducing apoptosis in human leukemia Jurkat T and breast cancer cells [96, 123]. Wu and Fang [124] reported that apigenin induced apoptotic cell death by its inhibitory properties on chymotrypsin-like and trypsin-like proteasome catalytic activities in four different human cancer cell lines.

3.1.7. Reactive Oxygen Species-Dependent and –Independent Apoptosis

Many recent studies have demonstrated the induction of apoptosis by apigenin through production of reactive oxygen species (ROS), in contrast to the described anti-oxidant/anti-inflammatory effects of apigenin. Apigenin was shown to induce apoptosis through accumulation of ROS in prostate cancer 22Rv1 cells, which was accompanied by rapid glutathione depletion, disruption of mitochondrial membrane potential, and cytosolic release of cytochrome *c* [46]. These were confirmed by pre-treatment with N-acetylcysteine (NAC), a strong anti-oxidant, or enzyme catalase, which could reverse apigenin-induced apoptosis in 22Rv1 cells. This was supported by another study showing that apigenin-induced apoptosis in HepG2 human hepatoma cells, through elevation of intracellular ROS, could be blocked by treatment with inhibitors of NADPH oxidase (diphenylene iodonium, apocynin, and neopterine) [125].

In contrast to the critical role of ROS in apigenin-induced apoptosis, several reports have also described ROS-independent apoptosis by apigenin. Vargo et al. [72] reported that treatment with apigenin resulted in increased ROS and phosphorylation of the MAPKs, p38, and extracellular signal-regulated kinases (ERK). However, inhibition of ROS, p38, or ERK failed to block apoptosis, suggesting that these cellular responses induced by apigenin are not essential for induction of apoptosis. In addition, Jayasooriya et al. [73] found that apigenin could increase the level of intracellular ROS, whereas pretreatment with antioxidants, NAC, or glutathione resulted in completely attenuated ROS generation in human leukemia cells. Surprisingly, these antioxidants did not promote recuperation from apigenin-induced cell death, suggesting that ROS is not essential for suppression of apigenin-mediated apoptosis. Taken together, these results suggest that ROS may play a role in the mechanism of apoptosis by apigenin in cancer cells; however, essential role of ROS in apoptosis by apigenin may be cell type-dependent.

3.1.8. Other Mechanisms of Apigenin-Mediated Apoptosis

There are other important mechanisms utilized by apigenin in exertion of its anticancer activity through apoptosis. Apigenin was found to inhibit cyclin-dependent kinase 9 (CDK9) and block phosphorylation of the carboxy-terminal domain (CTD) of RNA polymerase II (RNAPII) at Ser-2, which correlated with the reduction of Mcl-1 protein expression and induction of apoptosis in T-cell leukemic cell line CEM [126]. CDK9, the core component of the positive transcription elongation factor b, phosphorylates Ser-2 residues in the CTD of RNAPII, which is required for transcript elongation [127]. Thus, inhibition of CDK9 leads to reduced RNA synthesis and subsequently to rapid downregulation of the short-lived anti-apoptotic protein Mcl-1, resulting in apoptosis induction in cancer cells. By using a pull-down experiment, apigenin was shown to interact with CDK9, suggesting the involvement of CKD9 in the anticancer activity of apigenin [127]. Similarly, this flavone was also reported to inhibit CDK6 activity through direct binding on CDK6-cyclin D complex [128]. CDK6 is a member of the CDKs group, which are involved in regulation of the cell cycle, neuronal function, transcription, and apoptosis and activated only in association with a regulatory partner, D-type cyclins. Apigenin was found to compete with ATP for binding into the ATP binding pocket in CDK6; thus, the kinase activity of the CDK6/cyclin D complex is inhibited. However, the question of whether inhibitory role of apigenin on CDK6 is involved in apoptotic cell death remains to be studied.

Recently, Choudhury and colleagues [129] reported that apigenin exerts synergistic anticancer activity with curcumin by binding at different sites of tubulin. It was found that this flavone bound to depolymerized microtubules and inhibited reassembly of cold depolymerized microtubules of A549 cells. It inhibited polymerization of purified tubulin and the stoichiometry of tubulin–apigenin binding was 1:1 and the binding constant (K_d) was 6.08 + 0.096 μM. Furthermore, the synergistic activity of apigenin and curcumin was also apparent from their strong depolymerizing effects on interphase microtubules and inhibitory effect of reassembly of cold depolymerized microtubules when used in combination, indicating that these ligands bind to tubulin at different sites. The combination of apigenin and curcumin caused apoptosis synergistically through induction of p53 expression and increase of Bax/Bcl-2 ratio, suggesting that disruption of microtubule networks by direct binding to tubulin might be involved in apoptotic effect of apigenin-curcumin combination.

Gonzalez-Mejia et al. [130] reported that apigenin-induced apoptosis of leukemia cells is mediated by residue-specific phosphorylation of heat-shock protein-27 (Hsp27). They observed that apigenin induced bimodal phosphorylation of Hsp27. Phosphorylation at early times was regulated by p38 whereas mediated by p38 activity and for some residues depended on PKCδ at a later time. In addition, overexpression of Hsp27 was found to reduce apigenin-induced apoptosis; however, expression of phosphor-mimic Hsp27-D significantly increased the susceptibility of leukemia cells to apigenin-induced apoptosis. The effect of apigenin in regulation of Hsp27 and its role in apoptosis illustrate its potential to circumvent Hsp27-mediated resistance of cancer cells. Overall, it is likely that apigenin triggers apoptosis through several modes of action. Although a number of studies have demonstrated that apigenin is non-toxic to normal human cells, a comprehensive animal model is warranted for study of the toxicological properties of apigenin in order to identify potential adverse effects.

3.2. Effect of Apigenin on Autophagy

3.2.1. Autophagy

Half a century ago, Christian de Duve coined the term "autophagy" ("self-eating", from the Greek words, auto "self" and phagein "to eat") to describe a process where the cell digests its cytoplasmic materials within lysosomes [131]. Nowadays, autophagy is defined as a cellular pathway by which cytoplasmic macromolecules and organelles are delivered to lysosomes for degradation [132]. Autophagy was first described in the late 1960s, however, the last decade has seen an explosion of interest and the linkage of autophagy to various diseases and various stresses, as well as normal developmental processes and aging [133-135].

To date, three forms of autophagy have been characterized depending on their function and the mode of delivery of the cargo to lysosomes [132]. Macroautophagy involves the formation of double membrane vesicles that engulf proteins and organelles such as mitochondria and peroxisomes [136]. These vesicles, called autophagosomes, ultimately fuse with lysosomes and the lysosomal hydrolases then degrade the contents in lysosomes. In microautophagy, lysosomes can also directly engulf cytoplasm by invagination, protrusion, and/or septation of the lysosomal limiting membrane. The second form of autophagy is chaperone-mediated autophagy (CMA) [137]. The protein to be amenable for lysosomal degradation via CMA contains an exposed pentapeptide motif (KFERQ), which is recognized by heat shock cognate 70 and then delivered directly to the lysosome via the lysosomal protein known as lysosome-associated membrane protein type 2A (LAMP2A). In CMA, internalization of the substrate protein is preceded by its unfolding, a step not required in the other types of autophagy. Finally, microautophagy, which is a non-selective lysosomal degradative process, involves direct engulfment of cytoplasmic cargo at a boundary membrane by autophagic tubes, which mediate both invagination and vesicle scission into the lumen [138]. Although several recent works have emphasized the important roles of microautophagy or CMA in tumor growth and progression, the best studied one is macroautophagy, and almost all of the work related to autophagy and cancer development, progression, and therapy refers to macroautophagy.

3.2.2. Autophagy and Cancer

A large body of evidence indicates that the relationship between autophagy and cancer is complex [139]. While autophagy can suppress tumor promotion, high levels of autophagy in multiple cancers indicate that autophagy primarily promotes the progression of established cancers [140-143]. However, recent research has shown that autophagy is more likely to be used as a tumor suppressor. For example, *beclin 1*$^{+/-}$ immortalized mouse mammary epithelial cells were more tumorigenic than their autophagy-competent counterparts in nude mice [144]. In agreement with these findings, exogenous expression of *beclin 1* into MCF7 breast cancer cells inhibited tumorigenicity and its overexpression inhibited tumor development [145, 146]. Other autophagy regulating genes such as *Atgs* and *Bif-1* also exhibit properties of tumor suppressors in mice [147-149]. Another important mechanism of autophagy is cancer preventive action by controlling the levels of selective autophagy adaptor proteins such as p62/SQSTM. The relationship between defective autophagy and p62/SQSTM1 accumulation with tumorigenesis was also shown when *p62/SQSTM1*$^{-/-}$ mice were resistant to Ras-induced lung carcinomas compared with wild-type animals, through a mechanism involving activation of the transcription factor nuclear factor (NF)-κB [150].

On the contrary, autophagy promotes tumor cell survival within the tumor microenvironment, due to *ras* oncogene-associated "autophagy addiction" [151]. Thus, while autophagy protects cells through the elimination of cellular waste, defective autophagy (in an apoptosis-deficient background) results in elevated genetic instability, chronic cell death and inflammation, thus promoting tumorigenesis [152], suggesting an interrelationship between autophagy and apoptosis in tissue homeostasis.

3.2.3. Autophagy and Cancer Therapy

Similar to its role in tumor progression, accumulated evidence implicates a biphasic role of autophagy following cancer therapeutics, with response increasing or diminishing their anticancer potential. On the one hand, conventional cytotoxic drugs (i.e., 5-fluorouracil, gemcitabine, camptothecin, temozolomide, imatinib, tamoxifen, cetuximab, and vorinostat) and irradiation have been shown to induce autophagy. It is activated as a protective mechanism to mediate the acquired resistance phenotype of some cancer cells' response to chemotherapy. On the other hand, autophagy may also play a role as a death executioner in induction of autophagic cell death, a form of physiological cell death which is contradictory to apoptosis [153]. Based on current genetic and pharmacological studies, it appears that anticancer drugs induce different effects of autophagy on cell survival in different cancer types [154].

This is of particular significance because several cancer therapeutics, including tyrosine kinase inhibitors, mammalian target of rapamycin (mTOR) inhibitors, ionizing radiation, microtubule inhibitors, and proteasome inhibitors, are known to modulate autophagy, which could interfere with tumor response [155]. The first reported clinical trial combining autophagy inhibitor with cancer therapy was conducted in glioblastoma patients treated with conventional radiation and carmustine therapy with or without chloroquine (CQ) [156]. Although this study was not adequately powered to detect a significant difference in survival, CQ increased overall survival from 11 months in the placebo arm to 24 months in the CQ arm. In contrast to the sensitization of tumors to conventional chemotherapeutics by inhibiting autophagy, the sensitivity of tumors to radiation has been shown to be enhanced by induction of autophagy with the mTOR inhibitor everolimus (RAD001) and further enhanced by combined inhibition of apoptosis and induction of autophagy [157, 158]. Therefore, conduct of further clinical trials is still necessary in order to clarify the benefits of chemotherapy in combination with autophagy inducers or autophagy inhibitors. The National Cancer Institute has several active clinical trials currently underway to investigate the additional benefit of autophagy modulation in cancer therapy (http://www.clinicaltrials.gov). The outcomes of these aforementioned preclinical as well as clinical studies provide insight into the potentials of the combination of cancer therapies targeting both apoptosis and autophagy, although the question of whether they counteract or synergize with each other remains a puzzle.

3.2.4. Induction of Autophagy by Apigenin

Several studies have indicated that apigenin induces autophagy in various cancer types [59, 64, 74, 159, 160]. The effect of apigenin on autophagy in cancer cells was first investigated by Ruela-de-Sousa and colleagues [74], who reported that apigenin initiated autophagy in TF1 erythroleukemia cells without triggering apoptosis. Exposure of TF1 cells to apigenin resulted in inhibition of mTOR, a main autophagy repressor, and its downstream p70S6K. Although no changes in beclin 1 levels and strong reduction of Atg 5, 7, and 12 were observed, apigenin induced non-electron-dense vacuoles and double-membrane vacuoles, strong evidence of autophagy in TF1 cells. These results suggested that autophagy is a dominant response to apigenin treatment in at least some types of leukemia. In addition, apigenin has been reported to exhibit autophagy-inducing effects in breast cancer T47D and MDA-MB-231 cells [59].

Apigenin caused autophagy, as characterized by the appearance of autophagosomes, accumulation of acidic vesicular organelles (AVOs), and escalation of the level of LC3-II, the processed form of LC3-I. The same study also reported that treatment with the autophagy inhibitor, 3-methyladenine (3-MA), significantly enhanced the apoptosis induced by apigenin, which was accompanied by an increase in the level of PARP cleavage. This was supported by another recent study by our group showing that apigenin induces both apoptosis and autophagy in HCT116 human colon cancer cells [64]. After treatment with apigenin, levels of LC3-II, redistribution of LC3 protein into cytoplasmic puncta (a characteristic of autophagy), and accumulation of AVO were increased in HCT116 cells, indicating that this flavone may affect autophagosome formation. Interestingly, treatment with 3-MA resulted in reversal of LC3 puncta-induced by apigenin; however, apigenin-mediated apoptotic cell death was enhanced. These results suggest that apigenin simultaneously induced apoptosis and autophagy; in addition, apigenin-mediated autophagy might play a cytoprotective role in apigenin-induced apoptosis.

Tong et al. [159], who investigated the effects of apigenin on AMPK modulation in human keratinocytes, found that apigenin-induced AMPK activation inhibited mTOR signaling and further induced autophagy in human keratinocytes, thereby suggesting that apigenin can act as a potent chemopreventive agent. In a study by Jeremic et al. [160], apigenin was shown to mediate autophagy in glioma C6 cells, as evident by increase of cytoplasmic acidic vesicles, induction of autophagosome-associated LC3-II and pro-autophagic protein beclin 1 content and decline of p62. Despite upregulation of autophagy, drastic activation of mTOR accompanied by moderate activation of its upstream activator Akt and decline of phosphorylation of AMPK were also detected, suggesting that the autophagy induced by apigenin was mTOR-independent.

3.2.5. Effect of Apigenin on Balance Between Apoptosis and Autophagy

Both self-destructive processes, apoptosis (self-killing) and autophagy (self-eating), are well-controlled processes that play essential roles in development, tissue homeostasis, and disease. It is noteworthy that autophagy is often observed in dying cells. In many circumstances, this represents an attempt of the cell to mitigate the stress before resorting to the irreversible and final solution of apoptosis. Activation of autophagy, in some other settings, might reflect crosstalk between apoptosis and autophagy. Indeed, autophagy is induced by many stresses such as organelle dysfunction, metabolic stress and pathogen infection, which eventually lead to apoptosis. In general, cells restore autophagy to baseline levels and return to their initial state after the stress is resolved. However, if the stress persists and autophagy fails to support cell survival any longer, cells might respond by activating apoptosis in order to ensure controlled and efficient elimination, without triggering local inflammation. Therefore, it is plausible that interplay between autophagy and apoptosis conferred an evolutionary advantage to cells. Moreover, while autophagy and apoptosis are distinct processes with fundamentally different molecular features, the involvement of several proteins (i.e., beclin 1, Bcl-2, Bcl-xL, mTOR, Atg5, p53, etc) that regulate both processes been reported [161-163].

During the past decade, the inhibition of autophagy by use of pharmacological inhibitors has been evaluated for augmentation of anticancer activity of conventional or targeted chemotherapeutics. Among numerous methods of autophagy inhibition, lysosomotropic agents including CQ, and its derivative HCQ, are the most studied and currently commercially available for clinical trials [164, 165]. A number of reports have indicated that natural products regulate autophagy in cancer cells in both positive and negative manners. Therefore, the cytoprotective autophagy induced by natural products was always present with apoptosis and this autophagy could be much more straightforward as the process was naturally regarded as cellular stress clearance. The cytoprotective autophagy is usually determined by using either pharmacologic inhibitors of autophagy (such as CQ, bafilomycin, 3-MA or ammonium chloride) or genetic silencing or knockdown of autophagy-associated genes (such as beclin 1, Atg 5, 7, or 12) to increase cancer cell sensitivity to autophagy-inducing stimuli, *via* promotion of apoptosis. Cao et al. [59] found that natural agent apigenin could induce both apoptosis and autophagy in breast cancer cells, therefore inhibition of autophagy with inhibitor 3-MA potentiated cell death. Another study from our group also reported similar results showing the chemopreventive effect of apigenin against colon cancer

[64]. Apigenin induced both apoptosis and autophagy in HCT116 cells. In addition, autophagy was inhibited by introduction of 3-MA enhanced apigenin-induced apoptosis in colon cancer cells, indicating a survival function of autophagy induced by apigenin. These results corroborate the line of evidence demonstrating that apigenin-induced autophagy, implicated in cell survival, contributes to the cytoprotective role of autophagy.

The other cytoprotective way by apigenin-induced autophagy in cancer cells is also available. The fact that autophagy has a dual role in cancer cells in that it can either help cells to survive during stress or end in apoptosis in the absence of cell rescue is well-accepted by researchers. Ruela-de-Sousa et al. [74], who examined the molecular mode of action in leukemia cells, found that apigenin-induced autophagy is not followed by apoptosis in TF1 cells. When TF cells were co-treated with apigenin and vincristine, one of the most common chemotherapeutics, cells became more resistant to vincristine-induced cell death. The influence of autophagy acquired by apigenin on cell death, particularly in the above mentioned cases, may be too complicated to explain currently, however, it may indicate that cytoprotective autophagy, as a decision-maker, may locate at the center of alternative courses of death or life of cancer cell fate.

While most studies focused on autophagy induction by apigenin, there is one study available on the inhibitory property of apigenin on autophagy. Mohan et al. [166] reported that apigenin could suppress autophagy in serum-starved human malignant neuroblastoma cells. This flavone suppressed the expression of beclin 1, an essential autophagy *protein.* Combination of apigenin with synthetic retinoid N-(4-hydroxyphenyl) retinamide (4-HPR) resulted in downregulation of autophagy inducing proteins including beclin 1, LC3 II, Toll-like receptor-4, and myeloid differentiation primary response 88 (Myd88) and in upregulation of autophagy inhibitory p-Akt/mTOR signaling pathway. Finally, combination of 4-HPR and apigenin worked synergistically to suppress autophagy and promote apoptosis in human malignant neuroblastoma cells.

CONCLUSION

Overall, this chapter clearly describes the beneficial effects of apigenin in prevention and treatment of cancer through multiple cell-signaling pathways including apoptosis and autophagy. Both *in vitro* and *in vivo* data from published literature so far suggest that apigenin can affect apoptosis and/or autophagy, which play a pivotal role in both promotion and suppression of carcinogenesis. However, in the future more detailed investigations are needed in order to completely understand its mechanisms of action and chemopreventive/therapeutic potential against human cancers. Interestingly, apigenin has been found to be bioavailable following oral administration in rats and mice, although no data are available for its pharmacokinetic and pharmacodynamic profiles in humans. The above mentioned evidence regarding apigenin is fascinating and warrants greater attention. Since cancer chemoprevention aims to stop the carcinogenic process or to prolong the onset of carcinogenesis by intervening with efficacious, non-toxic, and inexpensive agents to prevent, suppress, or reverse the malignant transformation, apigenin may satisfy most of these requirements. However, before apigenin can be used for prevention and/or treatment of diseases including cancer in the clinic, several unresolved questions must be answered, including those about bioavailability and safety profile in human. Future studies in this direction will hopefully bring this fascinating flavone to the forefront of cancer therapeutics.

ACKNOWLEDGMENTS

This study was supported by the 2014 Post-Doc. Development Program of Pusan National University. This work was also supported by the National Research Foundation of Korea (NRF) grant funded by the Korea government (MSIP) (No. 2009-0083538) and the R&D Program of MOTIE/KEIT (10040391, Development of Functional Food Materials and Device for Prevention of Aging-associated Muscle Function Decrease).

REFERENCES

[1] Siegel, R; Ma, J; Zou, Z; Jemal, A. Cancer statistics, 2014. *CA Cancer J. Clin.*, 2014, 64, 9-29.

[2] Bray, F; Jemal, A; Grey, N; Ferlay, J; Forman, D. Global cancer transitions according to the Human Development Index (2008-2030): a population-based study. *Lancet Oncol.*, 2012, 13, 790-801.

[3] Lippman, SM; Hong, WK. Cancer prevention science and practice. *Cancer Res.*, 2002, 62, 5119-25.

[4] Boeing, H; Bechthold, A; Bub, A; Ellinger, S; Haller, D; Kroke, A; Leschik-Bonnet, E; Muller, MJ; Oberritter, H; Schulze, M; Stehle, P; Watzl, B. Critical review: vegetables and fruit in the prevention of chronic diseases. *Eur. J. Nutr.*, 2012, 51, 637-63.

[5] Wang, X; Ouyang, Y; Liu, J; Zhu, M; Zhao, G; Bao, W; Hu, FB. Fruit and vegetable consumption and mortality from all causes, cardiovascular disease, and cancer: systematic review and dose-response meta-analysis of prospective cohort studies. *BMJ*, 2014, 349, g4490.

[6] Grosso, G; Buscemi, S; Galvano, F; Mistretta, A; Marventano, S; La Vela, V; Drago, F; Gangi, S; Basile, F; Biondi, A. Mediterranean diet and cancer: epidemiological evidence and mechanism of selected aspects. *BMC Surg.*, 2013, 13 Suppl 2, S14.

[7] Garcia-Closas, R; Gonzalez, CA; Agudo, A; Riboli, E. Intake of specific carotenoids and flavonoids and the risk of gastric cancer in Spain. *Cancer Causes Control*, 1999, 10, 71-5.

[8] Rossi, M; Negri, E; Talamini, R; Bosetti, C; Parpinel, M; Gnagnarella, P; Franceschi, S; Dal Maso, L; Montella, M; Giacosa, A; La Vecchia, C. Flavonoids and colorectal cancer in Italy. *Cancer Epidemiol. Biomarkers Prev.*, 2006, 15, 1555-8.

[9] Bosetti, C; Spertini, L; Parpinel, M; Gnagnarella, P; Lagiou, P; Negri, E; Franceschi, S; Montella, M; Peterson, J; Dwyer, J; Giacosa, A; La Vecchia, C. Flavonoids and breast cancer risk in Italy. *Cancer Epidemiol. Biomarkers Prev.*, 2005, 14, 805-8.

[10] Gates, MA; Vitonis, AF; Tworoger, SS; Rosner, B; Titus-Ernstoff, L; Hankinson, SE; Cramer, DW. Flavonoid intake and ovarian cancer risk in a population-based case-control study. *Int. J. Cancer*, 2009, 124, 1918-25.

[11] Horn-Ross, PL; John, EM; Canchola, AJ; Stewart, SL; Lee, MM. Phytoestrogen intake and endometrial cancer risk. *J. Natl. Cancer Inst.*, 2003, 95, 1158-64.

[12] Frankenfeld, CL; Cerhan, JR; Cozen, W; Davis, S; Schenk, M; Morton, LM; Hartge, P; Ward, MH. Dietary flavonoid intake and non-Hodgkin lymphoma risk. *Am. J. Clin. Nutr.*, 2008, 87, 1439-45.

[13] Szent-Gyorgyi, A. Perspectives for the bioflavonoids. *Ann. N. Y. Acad. Sci.*, 1955, 61, 732-5.

[14] Beecher, GR. Overview of dietary flavonoids: nomenclature, occurrence and intake. *J. Nutr.*, 2003, 133, 3248S-54S.

[15] Srivastava, JK; Shankar, E; Gupta, S. Chamomile: A herbal medicine of the past with bright future. *Mol. Med. Rep.*, 2010, 3, 895-901.

[16] Farzaei, MH; Abbasabadi, Z; Ardekani, MR; Rahimi, R; Farzaei, F. Parsley: a review of ethnopharmacology, phytochemistry and biological activities. *J. Tradit. Chin. Med.*, 2013, 33, 815-26.

[17] Kawai, M; Hirano, T; Higa, S; Arimitsu, J; Maruta, M; Kuwahara, Y; Ohkawara, T; Hagihara, K; Yamadori, T; Shima, Y; Ogata, A; Kawase, I; Tanaka, T. Flavonoids and related compounds as anti-allergic substances. *Allergol. Int.*, 2007, 56, 113-23.

[18] Birt, DF; Walker, B; Tibbels, MG; Bresnick, E. Anti-mutagenesis and anti-promotion by apigenin, robinetin and indole-3-carbinol. *Carcinogenesis*, 1986, 7, 959-63.

[19] Mafuvadze, B; Liang, Y; Besch-Williford, C; Zhang, X; Hyder, SM. Apigenin induces apoptosis and blocks growth of medroxyprogesterone acetate-dependent BT-474 xenograft tumors. *Horm. Cancer*, 2012, 3, 160-71.

[20] Mafuvadze, B; Benakanakere, I; Lopez Perez, FR; Besch-Williford, C; Ellersieck, MR; Hyder, SM. Apigenin prevents development of medroxyprogesterone acetate-accelerated 7,12-dimethylbenz(*a*) anthracene-induced mammary tumors in Sprague-Dawley rats. *Cancer Prev. Res. (Phila).*, 2011, 4, 1316-24.

[21] Shao, H; Jing, K; Mahmoud, E; Huang, H; Fang, X; Yu, C. Apigenin sensitizes colon cancer cells to antitumor activity of ABT-263. *Mol. Cancer Ther.*, 2013, 12, 2640-50.

[22] Chunhua, L; Donglan, L; Xiuqiong, F; Lihua, Z; Qin, F; Yawei, L; Liang, Z; Ge, W; Linlin, J; Ping, Z; Kun, L; Xuegang, S. Apigenin up-regulates transgelin and inhibits invasion and migration of colorectal cancer through decreased phosphorylation of AKT. *J. Nutr. Biochem.*, 2013, 24, 1766-75.

[23] Wang, QR; Yao, XQ; Wen, G; Fan, Q; Li, YJ; Fu, XQ; Li, CK; Sun, XG. Apigenin suppresses the growth of colorectal cancer xenografts via phosphorylation and up-regulated FADD expression. *Oncol. Lett.*, 2011, 2, 43-7.

[24] Engelmann, C; Blot, E; Panis, Y; Bauer, S; Trochon, V; Nagy, HJ; Lu, H; Soria, C. Apigenin--strong cytostatic and anti-angiogenic action in vitro contrasted by lack of efficacy in vivo. *Phytomedicine*, 2002, 9, 489-95.

[25] Leonardi, T; Vanamala, J; Taddeo, SS; Davidson, LA; Murphy, ME; Patil, BS; Wang, N; Carroll, RJ; Chapkin, RS; Lupton, JR; Turner, ND. Apigenin and naringenin suppress colon carcinogenesis through the aberrant crypt stage in azoxymethane-treated rats. *Exp. Biol. Med. (Maywood)*, 2010, 235, 710-7.

[26] Tatsuta, A; Iishi, H; Baba, M; Yano, H; Murata, K; Mukai, M; Akedo, H. Suppression by apigenin of peritoneal metastasis of intestinal adenocarcinomas induced by azoxymethane in Wistar rats. *Clin. Exp. Metastasis*, 2000, 18, 657-62.

[27] Zhong, Y; Krisanapun, C; Lee, SH; Nualsanit, T; Sams, C; Peungvicha, P; Baek, SJ. Molecular targets of apigenin in colorectal cancer cells: involvement of p21, NAG-1 and p53. *Eur. J. Cancer*, 2010, 46, 3365-74.

[28] Budhraja, A; Gao, N; Zhang, Z; Son, YO; Cheng, S; Wang, X; Ding, S; Hitron, A; Chen, G; Luo, J; Shi, X. Apigenin induces apoptosis in human leukemia cells and exhibits anti-leukemic activity in vivo. *Mol. Cancer Ther.*, 2012, 11, 132-42.

[29] Liu, LZ; Fang, J; Zhou, Q; Hu, X; Shi, X; Jiang, BH. Apigenin inhibits expression of vascular endothelial growth factor and angiogenesis in human lung cancer cells: implication of chemoprevention of lung cancer. *Mol. Pharmacol.*, 2005, 68, 635-43.

[30] Cai, J; Zhao, XL; Liu, AW; Nian, H; Zhang, SH. Apigenin inhibits hepatoma cell growth through alteration of gene expression patterns. *Phytomedicine*, 2011, 18, 366-73.

[31] Jeyabal, PV; Syed, MB; Venkataraman, M; Sambandham, JK; Sakthisekaran, D. Apigenin inhibits oxidative stress-induced macromolecular damage in N-nitrosodiethylamine (NDEA)-induced hepatocellular carcinogenesis in Wistar albino rats. *Mol. Carcinog.*, 2005, 44, 11-20.

[32] Baldasquin-Caceres, B; Gomez-Garcia, FJ; Lopez-Jornet, P; Castillo-Sanchez, J; Vicente-Ortega, V. Chemopreventive potential of phenolic compounds in oral carcinogenesis. *Arch. Oral Biol.*, 2014, 59, 1101-7.

[33] Gomez-Garcia, FJ; Lopez-Jornet, MP; Alvarez-Sanchez, N; Castillo-Sanchez, J; Benavente-Garcia, O; Vicente Ortega, V. Effect of the phenolic compounds apigenin and carnosic acid on oral carcinogenesis in hamster induced by DMBA. *Oral Dis.*, 2013, 19, 279-86.

[34] Silvan, S; Manoharan, S. Apigenin prevents deregulation in the expression pattern of cell-proliferative, apoptotic, inflammatory and angiogenic markers during 7,12-dimethylbenz[a]anthracene-induced hamster buccal pouch carcinogenesis. *Arch. Oral Biol.*, 2013, 58, 94-101.

[35] Silvan, S; Manoharan, S; Baskaran, N; Anusuya, C; Karthikeyan, S; Prabhakar, MM. Chemopreventive potential of apigenin in 7,12-dimethylbenz(a)anthracene induced experimental oral carcinogenesis. *Eur. J. Pharmacol.*, 2011, 670, 571-7.

[36] Lin, CC; Chuang, YJ; Yu, CC; Yang, JS; Lu, CC; Chiang, JH; Lin, JP; Tang, NY; Huang, AC; Chung, JG. Apigenin induces apoptosis through mitochondrial dysfunction in U-2 OS human osteosarcoma cells and inhibits osteosarcoma xenograft tumor growth in vivo. *J. Agric. Food Chem.*, 2012, 60, 11395-402.

[37] Hu, XW; Meng, D; Fang, J. Apigenin inhibited migration and invasion of human ovarian cancer A2780 cells through focal adhesion kinase. *Carcinogenesis*, 2008, 29, 2369-76.

[38] Li, ZD; Hu, XW; Wang, YT; Fang, J. Apigenin inhibits proliferation of ovarian cancer A2780 cells through Id1. *FEBS Lett.*, 2009, 583, 1999-2003.

[39] Lee, SH; Ryu, JK; Lee, KY; Woo, SM; Park, JK; Yoo, JW; Kim, YT; Yoon, YB. Enhanced anti-tumor effect of combination therapy with gemcitabine and apigenin in pancreatic cancer. *Cancer Lett.*, 2008, 259, 39-49.

[40] King, JC; Lu, QY; Li, G; Moro, A; Takahashi, H; Chen, M; Go, VL; Reber, HA; Eibl, G; Hines, OJ. Evidence for activation of mutated p53 by apigenin in human pancreatic cancer. *Biochim. Biophys. Acta*, 2012, 1823, 593-604.

[41] Shukla, S; Fu, P; Gupta, S. Apigenin induces apoptosis by targeting inhibitor of apoptosis proteins and Ku70-Bax interaction in prostate cancer. *Apoptosis*, 2014, 19, 883-94.

[42] Pandey, M; Kaur, P; Shukla, S; Abbas, A; Fu, P; Gupta, S. Plant flavone apigenin inhibits HDAC and remodels chromatin to induce growth arrest and apoptosis in human prostate cancer cells: in vitro and in vivo study. *Mol. Carcinog.*, 2012, 51, 952-62.

[43] Shukla, S; Gupta, S. Apigenin suppresses insulin-like growth factor I receptor signaling in human prostate cancer: an in vitro and in vivo study. *Mol. Carcinog.*, 2009, 48, 243-52.

[44] Kaur, P; Shukla, S; Gupta, S. Plant flavonoid apigenin inactivates Akt to trigger apoptosis in human prostate cancer: an in vitro and in vivo study. *Carcinogenesis*, 2008, 29, 2210-7.

[45] Shukla, S; Gupta, S. Molecular targets for apigenin-induced cell cycle arrest and apoptosis in prostate cancer cell xenograft. *Mol. Cancer Ther.*, 2006, 5, 843-52.

[46] Shukla, S; Gupta, S. Apigenin-induced prostate cancer cell death is initiated by reactive oxygen species and p53 activation. *Free Radic. Biol. Med.*, 2008, 44, 1833-45.

[47] Shukla, S; Mishra, A; Fu, P; MacLennan, GT; Resnick, MI; Gupta, S. Up-regulation of insulin-like growth factor binding protein-3 by apigenin leads to growth inhibition and apoptosis of 22Rv1 xenograft in athymic nude mice. *FASEB J.*, 2005, 19, 2042-4.

[48] Shukla, S; Bhaskaran, N; Babcook, MA; Fu, P; Maclennan, GT; Gupta, S. Apigenin inhibits prostate cancer progression in TRAMP mice via targeting PI3K/Akt/FoxO pathway. *Carcinogenesis*, 2014, 35, 452-60.

[49] Caltagirone, S; Rossi, C; Poggi, A; Ranelletti, FO; Natali, PG; Brunetti, M; Aiello, FB; Piantelli, M. Flavonoids apigenin and quercetin inhibit melanoma growth and metastatic potential. *Int. J. Cancer*, 2000, 87, 595-600.

[50] Byun, S; Park, J; Lee, E; Lim, S; Yu, JG; Lee, SJ; Chen, H; Dong, Z; Lee, KW; Lee, HJ. Src kinase is a direct target of apigenin against UVB-induced skin inflammation. *Carcinogenesis*, 2013, 34, 397-405.

[51] Wei, H; Tye, L; Bresnick, E; Birt, DF. Inhibitory effect of apigenin, a plant flavonoid, on epidermal ornithine decarboxylase and skin tumor promotion in mice. *Cancer Res.*, 1990, 50, 499-502.

[52] Johnstone, RW; Ruefli, AA; Lowe, SW. Apoptosis: a link between cancer genetics and chemotherapy. *Cell*, 2002, 108, 153-64.

[53] Hanahan, D; Weinberg, RA. Hallmarks of cancer: the next generation. *Cell*, 2011, 144, 646-74.

[54] Ashkenazi, A. Targeting the extrinsic apoptosis pathway in cancer. *Cytokine Growth Factor Rev.*, 2008, 19, 325-31.

[55] Green, DR; Kroemer, G. The pathophysiology of mitochondrial cell death. *Science*, 2004, 305, 626-9.

[56] Zhu, Y; Mao, Y; Chen, H; Lin, Y; Hu, Z; Wu, J; Xu, X; Xu, X; Qin, J; Xie, L. Apigenin promotes apoptosis, inhibits invasion and induces cell cycle arrest of T24 human bladder cancer cells. *Cancer Cell Int.*, 2013, 13, 54.

[57] Choi, EJ; Kim, GH. Apigenin causes G(2)/M arrest associated with the modulation of p21(Cip1) and Cdc2 and activates p53-dependent apoptosis pathway in human breast cancer SK-BR-3 cells. *J. Nutr. Biochem.*, 2009, 20, 285-90.

[58] Seo, HS; Ju, JH; Jang, K; Shin, I. Induction of apoptotic cell death by phytoestrogens by up-regulating the levels of phospho-p53 and p21 in normal and malignant estrogen receptor alpha-negative breast cells. *Nutr. Res.*, 2011, 31, 139-46.

[59] Cao, X; Liu, B; Cao, W; Zhang, W; Zhang, F; Zhao, H; Meng, R; Zhang, L; Niu, R; Hao, X; Zhang, B. Autophagy inhibition enhances apigenin-induced apoptosis in human breast cancer cells. *Chin. J. Cancer Res.*, 2013, 25, 212-22.

[60] Seo, HS; Choi, HS; Kim, SR; Choi, YK; Woo, SM; Shin, I; Woo, JK; Park, SY; Shin, YC; Ko, SG. Apigenin induces apoptosis via extrinsic pathway, inducing p53 and inhibiting STAT3 and NFkappaB signaling in HER2-overexpressing breast cancer cells. *Mol. Cell. Biochem.*, 2012, 366, 319-34.

[61] Seo, HS; Ku, JM; Choi, HS; Woo, JK; Jang, BH; Shin, YC; Ko, SG. Induction of caspase-dependent apoptosis by apigenin by inhibiting STAT3 signaling in HER2-overexpressing MDA-MB-453 breast cancer cells. *Anticancer Res.*, 2014, 34, 2869-82.

[62] Choi, EJ; Kim, GH. Apigenin Induces Apoptosis through a Mitochondria/Caspase-Pathway in Human Breast Cancer MDA-MB-453 Cells. *J. Clin. Biochem. Nutr.*, 2009, 44, 260-5.

[63] Zheng, PW; Chiang, LC; Lin, CC. Apigenin induced apoptosis through p53-dependent pathway in human cervical carcinoma cells. *Life Sci.*, 2005, 76, 1367-79.

[64] Lee, Y; Sung, B; Kang, YJ; Kim, DH; Jang, JY; Hwang, SY; Kim, M; Lim, HS; Yoon, JH; Chung, HY; Kim, ND. Apigenin-induced apoptosis is enhanced by inhibition of autophagy formation in HCT116 human colon cancer cells. *Int. J. Oncol.*, 2014, 44, 1599-606.

[65] Wang, W; Heideman, L; Chung, CS; Pelling, JC; Koehler, KJ; Birt, DF. Cell-cycle arrest at G2/M and growth inhibition by apigenin in human colon carcinoma cell lines. *Mol. Carcinog.*, 2000, 28, 102-10.

[66] Zhang, Q; Zhao, XH; Wang, ZJ. Cytotoxicity of flavones and flavonols to a human esophageal squamous cell carcinoma cell line (KYSE-510) by induction of G2/M arrest and apoptosis. *Toxicol. In Vitro*, 2009, 23, 797-807.

[67] Chen, J; Chen, J; Li, Z; Liu, C; Yin, L. The apoptotic effect of apigenin on human gastric carcinoma cells through mitochondrial signal pathway. *Tumour Biol.*, 2014, 35, 7719-26.

[68] Chan, LP; Chou, TH; Ding, HY; Chen, PR; Chiang, FY; Kuo, PL; Liang, CH. Apigenin induces apoptosis via tumor necrosis factor receptor- and Bcl-2-mediated pathway and enhances susceptibility of head and neck squamous cell carcinoma to 5-fluorouracil and cisplatin. *Biochim. Biophys. Acta*, 2012, 1820, 1081-91.

[69] Das, S; Das, J; Samadder, A; Boujedaini, N; Khuda-Bukhsh, AR. Apigenin-induced apoptosis in A375 and A549 cells through selective action and dysfunction of mitochondria. *Exp. Biol. Med. (Maywood)*, 2012, 237, 1433-48.

[70] Das, S; Das, J; Samadder, A; Paul, A; Khuda-Bukhsh, AR. Strategic formulation of apigenin-loaded PLGA nanoparticles for intracellular trafficking, DNA targeting and improved therapeutic effects in skin melanoma in vitro. *Toxicol. Lett.*, 2013, 223, 124-38.

[71] Arango, D; Parihar, A; Villamena, FA; Wang, L; Freitas, MA; Grotewold, E; Doseff, AI. Apigenin induces DNA damage through the PKCdelta-dependent activation of ATM and H2AX causing down-regulation of genes involved in cell cycle control and DNA repair. *Biochem. Pharmacol.*, 2012, 84, 1571-80.

[72] Vargo, MA; Voss, OH; Poustka, F; Cardounel, AJ; Grotewold, E; Doseff, AI. Apigenin-induced-apoptosis is mediated by the activation of PKCdelta and caspases in leukemia cells. *Biochem. Pharmacol.*, 2006, 72, 681-92.

[73] Jayasooriya, RG; Kang, SH; Kang, CH; Choi, YH; Moon, DO; Hyun, JW; Chang, WY; Kim, GY. Apigenin decreases cell viability and telomerase activity in human leukemia cell lines. *Food Chem. Toxicol.*, 2012, 50, 2605-11.

[74] Ruela-de-Sousa, RR; Fuhler, GM; Blom, N; Ferreira, CV; Aoyama, H; Peppelenbosch, MP. Cytotoxicity of apigenin on leukemia cell lines: implications for prevention and therapy. *Cell Death Dis.*, 2010, 1, e19.

[75] Landis-Piwowar, KR; Milacic, V; Dou, QP. Relationship between the methylation status of dietary flavonoids and their growth-inhibitory and apoptosis-inducing activities in human cancer cells. *J. Cell. Biochem.*, 2008, 105, 514-23.

[76] Kim, BR; Jeon, YK; Nam, MJ. A mechanism of apigenin-induced apoptosis is potentially related to anti-angiogenesis and anti-migration in human hepatocellular carcinoma cells. *Food Chem. Toxicol.*, 2011, 49, 1626-32.

[77] Kim, KC; Choi, EH; Lee, C. Axl receptor tyrosine kinase is a novel target of apigenin for the inhibition of cell proliferation. *Int. J. Mol. Med.*, 2014, 34, 592-8.

[78] Lu, HF; Chie, YJ; Yang, MS; Lee, CS; Fu, JJ; Yang, JS; Tan, TW; Wu, SH; Ma, YS; Ip, SW; Chung, JG. Apigenin induces caspase-dependent apoptosis in human lung cancer A549 cells through Bax- and Bcl-2-triggered mitochondrial pathway. *Int. J. Oncol.*, 2010, 36, 1477-84.

[79] Pan, X; Yang, Z; Yang, Z; Zhou, S; Zhang, H; Zang, L. [Effect of apigenin on proliferation and apoptosis of human lung cancer NCI-H460 cells]. *Nan Fang Yi Ke Da Xue Xue Bao*, 2013, 33, 1137-40.

[80] Lu, HF; Chie, YJ; Yang, MS; Lu, KW; Fu, JJ; Yang, JS; Chen, HY; Hsia, TC; Ma, CY; Ip, SW; Chung, JG. Apigenin induces apoptosis in human lung cancer H460 cells through caspase- and mitochondria-dependent pathways. *Hum. Exp. Toxicol.*, 2011, 30, 1053-61.

[81] Hussain, AR; Khan, AS; Ahmed, SO; Ahmed, M; Platanias, LC; Al Kuraya, KS; Uddin, S. Apigenin induces apoptosis via downregulation of S-phase kinase-associated protein 2-mediated induction of p27Kip1 in primary effusion lymphoma cells. *Cell Prolif.*, 2010, 43, 170-83.

[82] Zhao, M; Ma, J; Zhu, HY; Zhang, XH; Du, ZY; Xu, YJ; Yu, XD. Apigenin inhibits proliferation and induces apoptosis in human multiple myeloma cells through targeting the trinity of CK2, Cdc37 and Hsp90. *Mol. Cancer*, 2011, 10, 104.

[83] Hossain, MM; Banik, NL; Ray, SK. N-Myc knockdown and apigenin treatment controlled growth of malignant neuroblastoma cells having N-Myc amplification. *Gene*, 2013, 529, 27-36.

[84] Torkin, R; Lavoie, JF; Kaplan, DR; Yeger, H. Induction of caspase-dependent, p53-mediated apoptosis by apigenin in human neuroblastoma. *Mol. Cancer Ther.*, 2005, 4, 1-11.

[85] Maggioni, D; Garavello, W; Rigolio, R; Pignataro, L; Gaini, R; Nicolini, G. Apigenin impairs oral squamous cell carcinoma growth in vitro inducing cell cycle arrest and apoptosis. *Int. J. Oncol.*, 2013, 43, 1675-82.

[86] Johnson, JL; de Mejia, EG. Flavonoid apigenin modified gene expression associated with inflammation and cancer and induced apoptosis in human pancreatic cancer cells

through inhibition of GSK-3beta/NF-kappaB signaling cascade. *Mol. Nutr. Food Res.*, 2013, 57, 2112-27.

[87] Hamacher, R; Saur, D; Fritsch, R; Reichert, M; Schmid, RM; Schneider, G. Casein kinase II inhibition induces apoptosis in pancreatic cancer cells. *Oncol. Rep.*, 2007, 18, 695-701.

[88] Morrissey, C; O'Neill, A; Spengler, B; Christoffel, V; Fitzpatrick, JM; Watson, RW. Apigenin drives the production of reactive oxygen species and initiates a mitochondrial mediated cell death pathway in prostate epithelial cells. *Prostate*, 2005, 63, 131-42.

[89] Gupta, S; Afaq, F; Mukhtar, H. Involvement of nuclear factor-kappa B, Bax and Bcl-2 in induction of cell cycle arrest and apoptosis by apigenin in human prostate carcinoma cells. *Oncogene*, 2002, 21, 3727-38.

[90] Oishi, M; Iizumi, Y; Taniguchi, T; Goi, W; Miki, T; Sakai, T. Apigenin sensitizes prostate cancer cells to Apo2L/TRAIL by targeting adenine nucleotide translocase-2. *PLoS ONE*, 2013, 8, e55922.

[91] Kim, SH; Kang, JG; Kim, CS; Ihm, SH; Choi, MG; Yoo, HJ; Lee, SJ. Apigenin induces c-Myc-mediated apoptosis in FRO anaplastic thyroid carcinoma cells. *Mol. Cell. Endocrinol.*, 2013, 369, 130-9.

[92] McIlwain, DR; Berger, T; Mak, TW. Caspase functions in cell death and disease. *Cold Spring Harb. Perspect. Biol.*, 2013, 5, a008656.

[93] Okun, I; Balakin, KV; Tkachenko, SE; Ivachtchenko, AV. Caspase activity modulators as anticancer agents. *Anticancer Agents Med. Chem.*, 2008, 8, 322-41.

[94] Masuelli, L; Marzocchella, L; Quaranta, A; Palumbo, C; Pompa, G; Izzi, V; Canini, A; Modesti, A; Galvano, F; Bei, R. Apigenin induces apoptosis and impairs head and neck carcinomas EGFR/ErbB2 signaling. *Front. Biosci. (Landmark Ed)*, 2011, 16, 1060-8.

[95] Way, TD; Kao, MC; Lin, JK. Apigenin induces apoptosis through proteasomal degradation of HER2/neu in HER2/neu-overexpressing breast cancer cells via the phosphatidylinositol 3-kinase/Akt-dependent pathway. *J. Biol. Chem.*, 2004, 279, 4479-89.

[96] Chen, D; Landis-Piwowar, KR; Chen, MS; Dou, QP. Inhibition of proteasome activity by the dietary flavonoid apigenin is associated with growth inhibition in cultured breast cancer cells and xenografts. *Breast Cancer Res.*, 2007, 9, R80.

[97] Das, S; Das, J; Samadder, A; Paul, A; Khuda-Bukhsh, AR. Efficacy of PLGA-loaded apigenin nanoparticles in Benzo[a]pyrene and ultraviolet-B induced skin cancer of mice: mitochondria mediated apoptotic signalling cascades. *Food Chem. Toxicol.*, 2013, 62, 670-80.

[98] Bishop, JM. Retroviruses and cancer genes. *Adv. Cancer Res.*, 1982, 37, 1-32.

[99] Hoffman, B; Liebermann, DA. *Oncogene*, 2008, 27, 6462-72.

[100] Yin, F; Giuliano, AE; Van Herle, AJ. Signal pathways involved in apigenin inhibition of growth and induction of apoptosis of human anaplastic thyroid cancer cells (ARO). *Anticancer Res.*, 1999, 19, 4297-303.

[101] Shukla, S; MacLennan, GT; Flask, CA; Fu, P; Mishra, A; Resnick, MI; Gupta, S. Blockade of beta-catenin signaling by plant flavonoid apigenin suppresses prostate carcinogenesis in TRAMP mice. *Cancer Res.*, 2007, 67, 6925-35.

[102] Reinhardt, HC; Schumacher, B. The p53 network: cellular and systemic DNA damage responses in aging and cancer. *Trends Genet.*, 2012, 28, 128-36.

[103] Vidal, A; Koff, A. Cell-cycle inhibitors: three families united by a common cause. *Gene*, 2000, 247, 1-15.

[104] Laptenko, O; Prives, C. Transcriptional regulation by p53: one protein, many possibilities. *Cell Death Differ.*, 2006, 13, 951-61.

[105] Vogelstein, B; Lane, D; Levine, AJ. Surfing the p53 network. *Nature*, 2000, 408, 307-10.

[106] Hoffman, WH; Biade, S; Zilfou, JT; Chen, J; Murphy, M. Transcriptional repression of the anti-apoptotic survivin gene by wild type p53. *J. Biol. Chem.*, 2002, 277, 3247-57.

[107] Riley, T; Sontag, E; Chen, P; Levine, A. Transcriptional control of human p53-regulated genes. *Nat. Rev. Mol. Cell Biol.*, 2008, 9, 402-12.

[108] Yu, J; Zhang, L; Hwang, PM; Rago, C; Kinzler, KW; Vogelstein, B. Identification and classification of p53-regulated genes. *Proc. Natl. Acad. Sci. USA,* 1999, 96, 14517-22.

[109] Chiang, LC; Ng, LT; Lin, IC; Kuo, PL; Lin, CC. Anti-proliferative effect of apigenin and its apoptotic induction in human Hep G2 cells. *Cancer Lett.*, 2006, 237, 207-14.

[110] Yang, SH; Liao, PH; Pan, YF; Chen, SL; Chou, SS; Chou, MY. The novel p53-dependent metastatic and apoptotic pathway induced by vitexin in human oral cancer OC2 cells. *Phytother. Res.*, 2013, 27, 1154-61.

[111] Bode, AM; Dong, Z. Post-translational modification of p53 in tumorigenesis. *Nat. Rev. Cancer*, 2004, 4, 793-805.

[112] Unger, T; Sionov, RV; Moallem, E; Yee, CL; Howley, PM; Oren, M; Haupt, Y. Mutations in serines 15 and 20 of human p53 impair its apoptotic activity. *Oncogene*, 1999, 18, 3205-12.

[113] Cai, X; Liu, X. Inhibition of Thr-55 phosphorylation restores p53 nuclear localization and sensitizes cancer cells to DNA damage. *Proc. Natl. Acad. Sci. USA*, 2008, 105, 16958-63.

[114] Navone, NM; Troncoso, P; Pisters, LL; Goodrow, TL; Palmer, JL; Nichols, WW; von Eschenbach, AC; Conti, CJ. p53 protein accumulation and gene mutation in the progression of human prostate carcinoma. *J. Natl. Cancer Inst.*, 1993, 85, 1657-69.

[115] Zhang, Q; Zhao, XH; Wang, ZJ. Flavones and flavonols exert cytotoxic effects on a human oesophageal adenocarcinoma cell line (OE33) by causing G2/M arrest and inducing apoptosis. *Food Chem. Toxicol.*, 2008, 46, 2042-53.

[116] Shukla, S; Gupta, S. Molecular mechanisms for apigenin-induced cell-cycle arrest and apoptosis of hormone refractory human prostate carcinoma DU145 cells. *Mol. Carcinog.*, 2004, 39, 114-26.

[117] Seo, YJ; Kim, BS; Chun, SY; Park, YK; Kang, KS; Kwon, TG. Apoptotic effects of genistein, biochanin-A and apigenin on LNCaP and PC-3 cells by p21 through transcriptional inhibition of polo-like kinase-1. *J. Korean Med. Sci.*, 2011, 26, 1489-94.

[118] Nalepa, G; Rolfe, M; Harper, JW. Drug discovery in the ubiquitin-proteasome system. *Nat. Rev. Drug Discov.*, 2006, 5, 596-613.

[119] Coux, O; Tanaka, K; Goldberg, AL. Structure and functions of the 20S and 26S proteasomes. *Annu. Rev. Biochem.*, 1996, 65, 801-47.

[120] Zhang, WG; Yu, JP; Wu, QM; Tong, Q; Li, SB; Wang, XH; Xie, GJ. Inhibitory effect of ubiquitin-proteasome pathway on proliferation of esophageal carcinoma cells. *World J. Gastroenterol.*, 2004, 10, 2779-84.

[121] Naujokat, C; Hoffmann, S. Role and function of the 26S proteasome in proliferation and apoptosis. *Lab. Invest.*, 2002, 82, 965-80.

[122] Orlowski, RZ. The role of the ubiquitin-proteasome pathway in apoptosis. *Cell Death Differ.*, 1999, 6, 303-13.
[123] Chen, D; Daniel, KG; Chen, MS; Kuhn, DJ; Landis-Piwowar, KR; Dou, QP. Dietary flavonoids as proteasome inhibitors and apoptosis inducers in human leukemia cells. *Biochem. Pharmacol.*, 2005, 69, 1421-32.
[124] Wu, YX; Fang, X. Apigenin, chrysin, and luteolin selectively inhibit chymotrypsin-like and trypsin-like proteasome catalytic activities in tumor cells. *Planta Med.*, 2010, 76, 128-32.
[125] Choi, SI; Jeong, CS; Cho, SY; Lee, YS. Mechanism of apoptosis induced by apigenin in HepG2 human hepatoma cells: involvement of reactive oxygen species generated by NADPH oxidase. *Arch. Pharm. Res.*, 2007, 30, 1328-35.
[126] Polier, G; Ding, J; Konkimalla, BV; Eick, D; Ribeiro, N; Kohler, R; Giaisi, M; Efferth, T; Desaubry, L; Krammer, PH; Li-Weber, M. Wogonin and related natural flavones are inhibitors of CDK9 that induce apoptosis in cancer cells by transcriptional suppression of Mcl-1. *Cell Death Dis.*, 2011, 2, e182.
[127] Wang, S; Fischer, PM. Cyclin-dependent kinase 9: a key transcriptional regulator and potential drug target in oncology, virology and cardiology. *Trends Pharmacol. Sci.*, 2008, 29, 302-13.
[128] Khuntawee, W; Rungrotmongkol, T; Hannongbua, S. Molecular dynamic behavior and binding affinity of flavonoid analogues to the cyclin dependent kinase 6/cyclin D complex. *J. Chem. Inf. Model.*, 2012, 52, 76-83.
[129] Choudhury, D; Ganguli, A; Dastidar, DG; Acharya, BR; Das, A; Chakrabarti, G. Apigenin shows synergistic anticancer activity with curcumin by binding at different sites of tubulin. *Biochimie*, 2013, 95, 1297-309.
[130] Gonzalez-Mejia, ME; Voss, OH; Murnan, EJ; Doseff, AI. Apigenin-induced apoptosis of leukemia cells is mediated by a bimodal and differentially regulated residue-specific phosphorylation of heat-shock protein-27. *Cell Death Dis.*, 2010, 1, e64.
[131] De Duve, C; Wattiaux, R. Functions of lysosomes. *Annu. Rev. Physiol.*, 1966, 28, 435-92.
[132] Jardon, MA; Rothe, K; Bortnik, S; Vezenkov, L; Jiang, X; Young, RN; Lum, JJ; Gorski, SM. Autophagy: from structure to metabolism to therapeutic regulation. *Autophagy*, 2013, 9, 2180-2.
[133] Rubinsztein, DC; Codogno, P; Levine, B. Autophagy modulation as a potential therapeutic target for diverse diseases. *Nat. Rev. Drug Discov.*, 2012, 11, 709-30.
[134] Rubinsztein, DC; Marino, G; Kroemer, G. Autophagy and aging. *Cell*, 2011, 146, 682-95.
[135] Kroemer, G; Marino, G; Levine, B. Autophagy and the integrated stress response. *Mol. Cell*, 2010, 40, 280-93.
[136] Ravikumar, B; Futter, M; Jahreiss, L; Korolchuk, VI; Lichtenberg, M; Luo, S; Massey, DC; Menzies, FM; Narayanan, U; Renna, M; Jimenez-Sanchez, M; Sarkar, S; Underwood, B; Winslow, A; Rubinsztein, DC. Mammalian macroautophagy at a glance. *J. Cell Sci.*, 2009, 122, 1707-11.
[137] Kaushik, S; Bandyopadhyay, U; Sridhar, S; Kiffin, R; Martinez-Vicente, M; Kon, M; Orenstein, SJ; Wong, E; Cuervo, AM. Chaperone-mediated autophagy at a glance. *J. Cell Sci.*, 2011, 124, 495-9.

[138] Mijaljica, D; Prescott, M; Devenish, RJ. Microautophagy in mammalian cells: revisiting a 40-year-old conundrum. *Autophagy*, 2011, 7, 673-82.
[139] White, E. Deconvoluting the context-dependent role for autophagy in cancer. *Nat. Rev. Cancer*, 2012, 12, 401-10.
[140] Lorin, S; Hamai, A; Mehrpour, M; Codogno, P. Autophagy regulation and its role in cancer. *Semin. Cancer Biol.*, 2013, 23, 361-79.
[141] Rosenfeldt, MT; O'Prey, J; Morton, JP; Nixon, C; MacKay, G; Mrowinska, A; Au, A; Rai, TS; Zheng, L; Ridgway, R; Adams, PD; Anderson, KI; Gottlieb, E; Sansom, OJ; Ryan, KM. p53 status determines the role of autophagy in pancreatic tumour development. *Nature*, 2013, 504, 296-300.
[142] Parkhitko, A; Myachina, F; Morrison, TA; Hindi, KM; Auricchio, N; Karbowniczek, M; Wu, JJ; Finkel, T; Kwiatkowski, DJ; Yu, JJ; Henske, EP. Tumorigenesis in tuberous sclerosis complex is autophagy and p62/sequestosome 1 (SQSTM1)-dependent. *Proc. Natl. Acad. Sci. USA*, 2011, 108, 12455-60.
[143] Rao, S; Tortola, L; Perlot, T; Wirnsberger, G; Novatchkova, M; Nitsch, R; Sykacek, P; Frank, L; Schramek, D; Komnenovic, V; Sigl, V; Aumayr, K; Schmauss, G; Fellner, N; Handschuh, S; Glosmann, M; Pasierbek, P; Schlederer, M; Resch, GP; Ma, Y; Yang, H; Popper, H; Kenner, L; Kroemer, G; Penninger, JM. A dual role for autophagy in a murine model of lung cancer. *Nat. Commun.*, 2014, 5, 3056.
[144] Karantza-Wadsworth, V; Patel, S; Kravchuk, O; Chen, G; Mathew, R; Jin, S; White, E. Autophagy mitigates metabolic stress and genome damage in mammary tumorigenesis. *Genes Dev.*, 2007, 21, 1621-35.
[145] Liang, XH; Jackson, S; Seaman, M; Brown, K; Kempkes, B; Hibshoosh, H; Levine, B. Induction of autophagy and inhibition of tumorigenesis by beclin 1. *Nature*, 1999, 402, 672-6.
[146] Kondo, Y; Kanzawa, T; Sawaya, R; Kondo, S. The role of autophagy in cancer development and response to therapy. *Nat. Rev. Cancer*, 2005, 5, 726-34.
[147] Takahashi, Y; Coppola, D; Matsushita, N; Cualing, HD; Sun, M; Sato, Y; Liang, C; Jung, JU; Cheng, JQ; Mule, JJ; Pledger, WJ; Wang, HG. Bif 1 interacts with Beclin 1 through UVRAG and regulates autophagy and tumorigenesis. *Nat. Cell Biol.*, 2007, 9, 1142-51.
[148] Yousefi, S; Perozzo, R; Schmid, I; Ziemiecki, A; Schaffner, T; Scapozza, L; Brunner, T; Simon, HU. Calpain-mediated cleavage of Atg5 switches autophagy to apoptosis. *Nat. Cell Biol.*, 2006, 8, 1124-32.
[149] Marino, G; Salvador-Montoliu, N; Fueyo, A; Knecht, E; Mizushima, N; Lopez-Otin, C. Tissue-specific autophagy alterations and increased tumorigenesis in mice deficient in Atg4C/autophagin-3. *J. Biol. Chem.*, 2007, 282, 18573-83.
[150] Duran, A; Linares, JF; Galvez, AS; Wikenheiser, K; Flores, JM; Diaz-Meco, MT; Moscat, J. The signaling adaptor p62 is an important NF-kappaB mediator in tumorigenesis. *Cancer Cell*, 2008, 13, 343-54.
[151] Mancias, JD; Kimmelman, AC. Targeting autophagy addiction in cancer. *Oncotarget*, 2011, 2, 1302-6.
[152] Liu, JJ; Lin, M; Yu, JY; Liu, B; Bao, JK. Targeting apoptotic and autophagic pathways for cancer therapeutics. *Cancer Lett.*, 2011, 300, 105-14.
[153] Bincoletto, C; Bechara, A; Pereira, GJ; Santos, CP; Antunes, F; Peixoto da-Silva, J; Muler, M; Gigli, RD; Monteforte, PT; Hirata, H; Jurkiewicz, A; Smaili, SS. Interplay

between apoptosis and autophagy, a challenging puzzle: new perspectives on antitumor chemotherapies. *Chem. Biol. Interact.*, 2013, 206, 279-88.

[154] Yang, ZJ; Chee, CE; Huang, S; Sinicrope, FA. The role of autophagy in cancer: therapeutic implications. *Mol. Cancer Ther.*, 2011, 10, 1533-41.

[155] Benbrook, DM; Long, A. Integration of autophagy, proteasomal degradation, unfolded protein response and apoptosis. *Exp. Oncol.*, 2012, 34, 286-97.

[156] Sotelo, J; Briceno, E; Lopez-Gonzalez, MA. Adding chloroquine to conventional treatment for glioblastoma multiforme: a randomized, double-blind, placebo-controlled trial. *Ann. Intern. Med.*, 2006, 144, 337-43.

[157] Cao, C; Subhawong, T; Albert, JM; Kim, KW; Geng, L; Sekhar, KR; Gi, YJ; Lu, B. Inhibition of mammalian target of rapamycin or apoptotic pathway induces autophagy and radiosensitizes PTEN null prostate cancer cells. *Cancer Res.*, 2006, 66, 10040-7.

[158] Kim, KW; Hwang, M; Moretti, L; Jaboin, JJ; Cha, YI; Lu, B. Autophagy upregulation by inhibitors of caspase-3 and mTOR enhances radiotherapy in a mouse model of lung cancer. *Autophagy*, 2008, 4, 659-68.

[159] Tong, X; Smith, KA; Pelling, JC. Apigenin, a chemopreventive bioflavonoid, induces AMP-activated protein kinase activation in human keratinocytes. *Mol. Carcinog.*, 2012, 51, 268-79.

[160] Jeremic, I; Tadic, V; Isakovic, A; Trajkovic, V; Markovic, I; Redzic, Z; Isakovic, A. The mechanisms of in vitro cytotoxicity of mountain tea, Sideritis scardica, against the C6 glioma cell line. *Planta Med.*, 2013, 79, 1516-24.

[161] Maiuri, MC; Zalckvar, E; Kimchi, A; Kroemer, G. Self-eating and self-killing: crosstalk between autophagy and apoptosis. *Nat. Rev. Mol. Cell Biol.*, 2007, 8, 741-52.

[162] El-Khattouti, A; Selimovic, D; Haikel, Y; Hassan, M. Crosstalk between apoptosis and autophagy: molecular mechanisms and therapeutic strategies in cancer. *J. Cell Death*, 2013, 6, 37-55.

[163] Booth, LA; Tavallai, S; Hamed, HA; Cruickshanks, N; Dent, P. The role of cell signalling in the crosstalk between autophagy and apoptosis. *Cell. Signal.*, 2014, 26, 549-55.

[164] Maes, H; Rubio, N; Garg, AD; Agostinis, P. Autophagy: shaping the tumor microenvironment and therapeutic response. *Trends Mol. Med.*, 2013, 19, 428-46.

[165] Jiang, P; Mizushima, N. Autophagy and human diseases. *Cell Res.*, 2014, 24, 69-79.

[166] Mohan, N; Banik, NL; Ray, SK. Combination of N-(4-hydroxyphenyl) retinamide and apigenin suppressed starvation-induced autophagy and promoted apoptosis in malignant neuroblastoma cells. *Neurosci. Lett.*, 2011, 502, 24-9.

In: Apigenin and Naringenin
Editor: Nilus M. Stacks
ISBN: 978-1-63463-987-3
© 2015 Nova Science Publishers, Inc.

Chapter 7

ROLE OF APIGENIN IN CANCER PREVENTION

Rit Bahadur Gurung, Ramesh Prasad Pandey and Jae Kyung Sohng*
Institute of Biomolecule Reconstruction (iBR),
Department of Pharmaceutical Engineering,
Sun Moon University, Republic of Korea

ABSTRACT

Apigenin (4',5,7-trihydoxyflavone, $C_{15}H_{10}O_5$) is one of the most bioactive naturally occurring flavonoids, found in fruits, plant-derived beverages, vegetables, wheat sprouts, and some seasonings. It has been found to possess anti-oxidant, anti-inflammatory, and anti-cancer properties. Research on the effects of apigenin at the cellular and molecular levels have shown positive results against breast, cervical, colon, lung, hematological, prostate, ovarian, thyroid, gastric, pancreatic, endometrial, liver, and skin cancers. Recent findings have also indicated that apigenin may be an alternative medication for prevention and/or therapy of allergic asthma and Alzheimer's disease. There is little evidence to date to suggest that apigenin promotes adverse metabolic effects when consumed as part of a normal diet. This chapter describes the natural sources, chemical structure, biological properties, and role of apigenin in cancer prevention. Several lines of evidence suggest that apigenin holds much potential to be developed as a cancer chemopreventive agent despite its low bioavailability.

INTRODUCTION

The flavonoids are a family of polyphenolic phytochemicals that are common constituents of the human diet. They occur naturally in a broad range of plants, including fruits, vegetables, and beverages, such as wine and tea. They are usually found in the form of glycosides and sometimes as acylglycosides, though acylated, methylated, and sulfated derivatives are less frequent and found in lower concentrations in fruits and vegetables (Ross

* Prof. Jae Kyung Sohng; E-mail: sohng@sunmoon.ac.kr; Fax: 82-41-544-2919; Tel: 82-41-530-2246.

and Kasum, 2002; Patel et al., 2007). Flavonoids have the C6-C3-C6 general structural backbone in which two C6 units (rings A and B) are phenolic in nature and a heterocyclic pyrane ring C connects the two phenolic rings. Based on the chemical structures, flavonoids are categorized mainly into flavones, isoflavones, flavonols, flavanols (flavan-3-ols or catechins), flavanonols, flavanones, neoflavonoids, anthocyanidins, and chalcones (Harborne and Williams, 2000) (Fig. 1). These structurally diverse flavonoids have been a rich source of compounds with a variety of biological activities (Mahomoodally et al., 2005). The activities of flavonoids are directly related to their chemical structures. The degree of hydroxylation, other substitutions and conjugations, and degree of polymerization greatly affect the chemical nature of flavonoids (Kelly et al., 2002). Presently, there is an upsurge of interest in flavonoids and their derivatives because several *in vitro* and *in vivo* results have shown that they exert numerous beneficial effects, which include anti-oxidant, anti-inflammatory, cardioprotective, anti-viral, anti-allergic, and anti-tumor activities (Ielpo et al., 2000; Toker et al., 2004; Shukla and Gupta, 2010).

Research results suggest foods rich in fruit and vegetables are linked with a reduced risk of common human cancers (Hertog et al., 1994; Chung et al., 2003, Bonfili et al., 2008; Tang et al., 2009; Aune et al., 2011; Aune et al., 2012; Reiss et al., 2012).

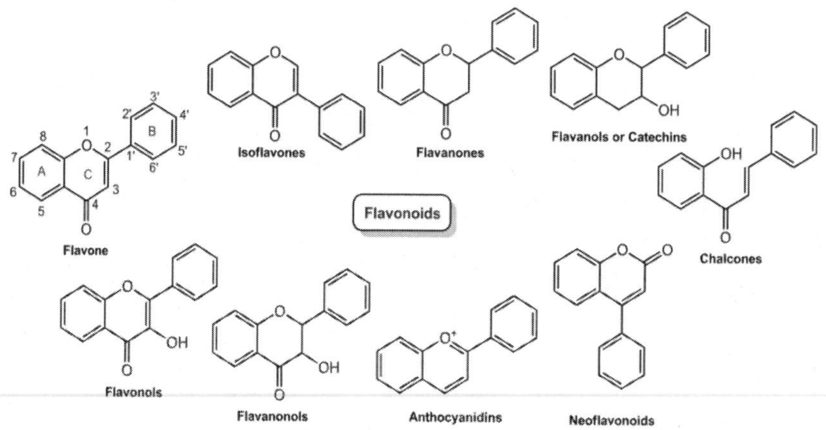

Figure 1. Classification of flavonoids and their basic chemical structures.

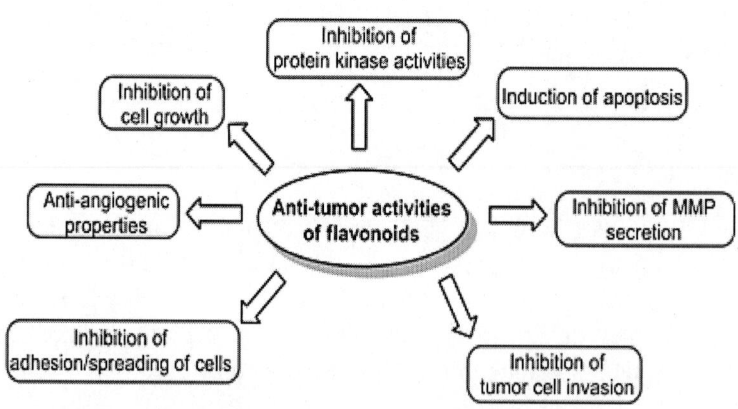

Figure 2. Antitumor activities of flavonoids.

As consumption of fruits and vegetables has been associated with a reduced risk of human cancers (Clere et al., 2011), dietary flavonoids and some other polyphenols have been identified as potential cancer-preventative components of fruits and vegetables (Zhang et al., 2009; Butler et al., 2010; Ni et al., 2012; Hui et al., 2012). These compounds have been proposed as primary chemopreventative agents, mainly due to their abilities to specifically inhibit tumor cell proliferation through obstruction of signal transduction, their anti-oxidant effects and/or their anti-metastatic properties, through inhibition of tumor angiogenesis (He et al., 2009; Pan et al., 2011; Zhao et al., 2009) (Fig. 2). Nevertheless, the specific mechanisms of action of most flavonoids in cancer prevention are not yet clear. However, numerous *in vitro* and animal model data suggest that flavonoids modulate important cellular and molecular mechanisms related to carcinogenesis (Clere et al., 2011).

Apigenin is a common dietary flavonoid, belonging to the flavone subclass of flavonoids and is present at high levels in several fruits and vegetables (Shukla and Gupta, 2010). It is known for a wide range of biological activities, including anti-inflammatory, anti-oxidant, anti-allergic, anti-osteoporotic, anti-viral, and anti-cancer effects (Kim et al., 2004; García-Lafuente et al., 2009; Park et al., 2008; Jayasooriya et al., 2012). It is one of the molecule that has gained substantial attention as valuable agent for human health with cancer preventive and/or therapeutic properties. Even at high doses, no toxicity was observed with apigenin (Hollman and Katan, 1999; Ross and Kasum, 2002).

With more than 10 million new cases every year, cancer has become one of the most devastating diseases worldwide. According to the World Health Organization (WHO), "cancer" is a generic term used for a group of diseases characterized by the uncontrolled growth and spread of abnormal cells, which can affect any part of the body. Cancer begins when normal cells change and grow uncontrollably, forming a mass called a tumor. A tumor can be cancerous or benign. A cancerous tumor is malignant, meaning it can spread to other parts of the body. A benign tumor means the tumor will not spread. The main defining feature of cancer is the rapid creation of abnormal cells that grow beyond their usual boundaries, and which can then invade adjoining parts of the body and spread to other organs. In the year 2012, there were 14.1 million new cancer cases, 8.2 million cancer deaths, and 32.6 million people living with cancer (Ferlay et al., 2013).

Cancer is caused by both external and internal factors. External factors include tobacco, chemicals, radiation, and infectious organisms whereas inherited mutations, hormones, immune conditions, and mutations that occur from metabolism are internal factors (American Cancer Society, 2014). Ageing is another fundamental factor for the development of cancer. Unhealthy diet and physical inactivity are also responsible for causing cancer. However, cancer can be reduced and controlled by implementing evidence-based strategies for cancer prevention, and the early detection of cancer and management of patients with cancer. According to one estimate, more than two-thirds of human cancers could be prevented by changes in lifestyle. Many cancers have a high chance of cure if detected early and treated adequately. One of the methods to prevent cancer uses overlapping strategies, such as chemoprevention or dietary cancer prevention.

The term 'chemoprevention' was first coined by Michael B. Sporn in 1976. Chemoprevention is an emerging discipline with great potential in the field of cancer treatment, defined as the use of a natural, synthetic, or biological agent to arrest or reverse the disease process of carcinogenesis at its very inception or during its early stages of progression towards invasive malignancy (Sporn, 1976). Epidemiological reports have suggested that dietary phytochemicals have the potential to act as chemopreventative agents. The use of dietary phytochemicals for chemoprevention is thought out to be an economical, acceptable, and accessible approach to cancer control and management (Surh, 2003).

NATURAL SOURCES, CHEMISTRY, AND PROPERTIES OF APIGENIN

Natural Sources

Apigenin is most commonly isolated from the plant *Matricaria chamomilla* (synonym: *Matricaria recutita*), a well-known medicinal plant species from the Asteraceae family, often referred to as the "star among medicinal species." (Singh et al., 2011). Other popular and ample sources include grapefruits, onions, oranges, parsley (Shukla and Gupta, 2010), chamomile tea (Singh et al., 2011; McKay and Blumberg, 2006). It is also found at high levels in celery, yarrow, tarragon, cilantro, passion flower, horehound, basil, foxglove, oregano, licorice, flax, coneflower (Birt et al., 2001; Patel et al., 2007), Chinese cabbage, bell pepper, garlic, and guava (Miean et al., 2001). It is also found in red wine (Bevilacqua et al., 2004), beer (Gerhäuser, 2005), wheat sprouts (Manach et al., 2004) and is an active ingredient in the memory herb *Ginkgo biloba* (Patel et al., 2007). Apigenin in diets and herbal sources is present mostly as glycoside conjugates and various acylated derivatives, which have higher water solubility than the free aglycone and remain mostly unchanged by various cooking methods (Hollman and Katan, 1999; Ross and Kasum, 2002; Patel et al., 2007; Gurung et al., 2013). The common apigenin glycosides are apiin, apigenin-7-*O*-glucoside (apigetrin), apigenin-8-C-glucoside (vitexin), apigenin-6-C-glucoside (isovitexin), and apigenin 7-*O*-neohesperidoside (rhoifolin) (Srivastava and Gupta 2007; da Silveira et al., 2010).

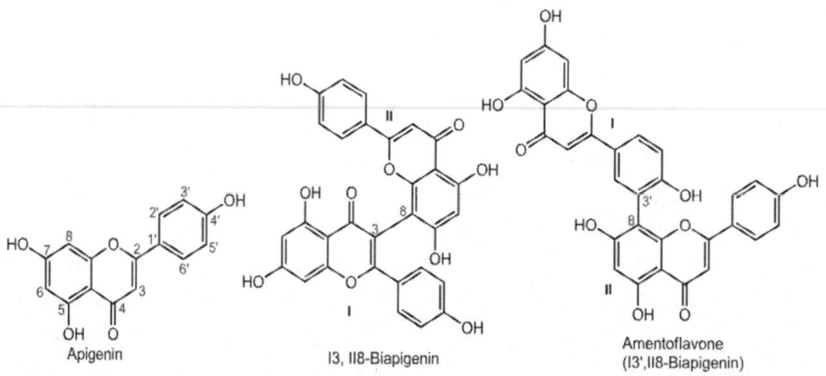

Figure 3. Chemical structure of apigenin and its dimers, I3, II8 Biapigenin and Amentoflavone.

Chemistry and Properties

Apigenin ($C_{15}H_{10}O_5$; 4',5,7-trihydroxyflavone) is a member of the flavone subclass of flavonoids. Its IUPAC name is 5,7-dihydroxy-2-(4-hydroxyphenyl) chromen-4-one and has molecular weight of 270.24. When pure, it forms yellow needles and has melting point of 347.5°C. It is practically insoluble in water (183 mg/L at 25°C), moderately soluble in hot alcohol, soluble in dimethyl sulfoxide, and dilute potassium hydroxide with an intense yellow color (Lide, 1997; Budavari, 1997). The UV max (ethanol) is 269, 340 nm (epsilon 18,800; 20,900) (O'Neil, 2006). Apigenin is also found in dimeric forms in plants. I3, II8-Biapigenin and amentoflavone (I3', II8-Biapigenin) are the dimers of apigenin, isolated from *Hypericum perforatum*, *Ginkgo biloba*, and *Xerophyta plicata* (Berghőfer and Hőlzl, 1987; Williams et al., 1987; Pan et al., 2005) (Fig. 3).

ROLE OF APIGENIN IN CANCER PREVENTION

Plants have played a dominant role in the development of sophisticated traditional medicine systems. According to WHO, ~80% of the population in some Asian and African countries depend on traditional medicine for primary healthcare (Khazir et al., 2014). Plant products also have a long history of use in the treatment of cancer. Hartwell in December 1967 published the first of a series of articles in *Lloydia* on the use of plants by humans against cancer in which he listed more than 3000 plant species having anti-cancer properties (Hartwell, 1967). Numerous phytochemicals derived from edible plants have been reported to interfere with specific stages of the carcinogenic process. In the past few years, several studies focused on phytochemicals and their health-related properties have shown the potential of dietary polyphenols as anti-carcinogenic agents (Watson et al., 2000; Yang et al., 2001).

Among the dietary polyphenols, apigenin has gained particular interest because of its low intrinsic toxicity and effects on normal versus cancer cells, compared with other structurally related flavonoids (Gupta et al., 2001). For centuries, the parts of plants or their extracts containing apigenin have been used in traditional or alternative medicine (Patel et al., 2007). Although attention has been paid on apigenin as an alternative anti-cancer compound, its mechanism of action remains unclear and apparently shows differences between different cancer types (Ruela-de-Sousa et al., 2010). Apigenin exerts cytotoxic effects on different cancer cell lines in which apoptosis plays a significant role (Kilani-Jaziri et al., 2012). The mechanism of apigenin-induced apoptosis include stimulation of gap junctional communication, inhibition of transformation, angiogenesis, and activation of nuclear transcription factor-kappa B (NF-κB) (Chan et al., 2012). Apigenin has also been reported to induce apoptosis through caspases-3, -8, and -9 (Iwasaki et al., 2009; Vargo et al., 2006; Lu et al., 2010). Apigenin treatment has been shown to induce G1 cell-cycle phase arrest and p53-dependent apoptosis, associated with a noticeable increase in the expression of p21/WAF1, and induction of Fas/APO-1 and caspase-3 expression. Additionally, apigenin decreases Bcl-2 expression (Zheng et al., 2005). Reversible G2/M cell-cycle arrest, linked with inhibition of p34 (cdc2) kinase and reduced accumulation of p34 (cdc2) and cyclin B1 proteins have been reported in colon carcinoma cell lines (Wang et al., 2000). Apigenin can also inhibit the estrogen receptor-α (ERα)-dependent pathway through downregulating protein levels of ERα and 'amplified in breast cancer-1' (AIB1) (Long et al., 2008). Furthermore, apigenin has been reported to increase the stability of the tumor suppressor protein, p53 in normal cells (Plaumann et al., 1996). Apigenin has been also shown to potentiate γ-aminobutyric acid type A ($GABA_A$) receptors, by binding to the benzodiazepine allosteric site (Viola et al., 1995). It also activates adenosine receptors (Jacobson et al., 2002), potentially inducing or facilitating sleep, and antagonizes N-methyl-D-aspartate (NMDA) receptors (Losi et al., 2004). In this chapter, we have attempted to summarize some major research works describing the role of apigenin in various cancers.

Breast Cancer

Breast cancer remains the most common malignancy and the predominant cause of cancer-related death in women (Are et al., 2013). It develops in the breast tissue, primarily in the milk ducts (ductal carcinoma) or glands (lobular carcinoma). The cancer is still called and treated as "breast cancer" even if it is first discovered after the cells have travelled to other areas of the body. A cancer of this type is referred to as metastatic or advanced breast cancer.

According to the American Cancer Society (ACS) in 2014, the estimated *in situ* cases, invasive cases, and deaths due to female breast cancer were 62,570, 232,670, and 40,000 respectively in the US alone. Similarly, about 2,360 men were estimated to be diagnosed with breast cancer and 430 of them will die from the disease (American Cancer Society, 2014).

Breast cancer is a hormone-dependent tumor because its development and growth can be dependent on the expression of estrogen receptors (ER) and androgen receptors, respectively. Recent findings suggest that low-dose apigenin has the potential to slow or prevent breast cancer progression. Harrison et al. has shown that a subcytotoxic concentration of apigenin inhibits DNA synthesis in a panel of human breast cancer cell lines (MDA-MB-213, MDA-MB-468, MCF-7, SK-BR-3). Decreased proliferation of MDA-MB-468 cells in the presence of apigenin was associated with G_2/M phase cell cycle arrest and the production of reactive oxygen species. Apigenin-treated MDA-MB-468 cells also showed reduced phosphorylation of Akt (protein kinase B), which is an essential effector serine/threonine kinase in the phosphatidylinositide 3-kinase pathway that promotes tumor growth and progression (Harrison et al., 2014).

Previous studies have shown that high concentrations of apigenin and other flavones induce apoptosis in breast cancer cells (Chen et al., 2007; Seelinger et al., 2008; Choi and Kim, 2009; Seo et al., 2012). Although apigenin has been found to target both estrogen receptor (ER)-dependent and -independent pathways, it seems to be somewhat more potent against ER-dependent human breast cancer cell lines (Sak, 2014). Mafuvadze et al. reported that the short-term administration of apigenin to Sprague Dawley rats by intraperitoneal injection protected against the development of medroxyprogesterone acetate (MPA)-accelerated 7,12-dimethylbenzen(a)anthracene (DMBA)-induced mammary tumors (Mafuvadze et al., 2011). The same group also found that intraperitoneal administration of apigenin effectively prevented progestin-dependent progression of human breast cancer cells (Mafuvadze et al., 2012). However, the route of administration may influence its action. Furthermore, caution should be exercised when considering apigenin as a dietary supplement for preventing hormone-dependent breast cancer (Mafuvadze et al., 2013).

Cervical Cancer

When the cancer starts in the cervix, it is called cervical cancer. The cervix is the lower, narrow end of the uterus, which connects the vagina to the upper part of the uterus. This is the second most common type of cancer in women at a global level. The ACS estimated that about 12,360 women will be diagnosed with cervical cancer in the United States in 2014 and 4,020 deaths will occur from the disease. Mortality rates have declined rapidly in past decades due especially to prevention and early detection as a result of screening with the Pap test (American Cancer Society, 2014).

In 2005, Zheng et al., for the first time, reported that apigenin inhibited the growth of human cervical carcinoma HeLa cells through an apoptotic pathway. Apigenin inhibited cell growth, caused G1 phase growth arrest, and induced apoptosis, which was p53-dependent and associated with a marked increase in the expression of p21/WAF1 protein and with the induction of Fas/APO-1 and caspase-3 expression. Apigenin also decreased the expression of Bcl-2 protein, an anti-apoptotic factor (Zheng et al., 2005). Similarly, another study demonstrated that apigenin can interfere with cell proliferation, cell survival, and gap junctional coupling. Exposure of non-invasive wild-type HeLa cells and their connexin43 (Cx43)-transfected counterparts to apigenin resulted in a significant and reversible inhibition of translocation of both cell types. The effect of apigenin on cell proliferation was less pronounced, especially at low apigenin concentration, whereas its influence on cell motility correlated with the reduction of the invasive potential of HeLa Cx43 cells (Czyz et al., 2005).

In a study carried out with medicinal plant feverfew (*Tanacetum parthenium* (L.) Schultz-Bip), an ethanolic extract containing parthenolide, camphor, luteolin, and apigenin showed anti-proliferative activity against two human breast cancer cell lines (Hs605T and MCF-7) and one human cervical cancer cell line (SiHa) (Wu et al., 2006). Recently, Liu et al. reported that apigenin inhibited the self-renewal capacity and the protein expression of the protein kinase casein kinase 2 (CK2) in HeLa-derived sphere-forming cells (SFCs; cells possessing properties similar to cervical cancer stem cells), in a dose-dependent manner (Liu et al., 2014). Apigenin is probably more potent in killing cervical cancer cells than cells of melanoma and lung cancer; the same seems also to be true for chrysin (Sak, 2014).

Colon and Rectum (Colorectal) Cancer

Colorectal cancer is the third most common cancer in both men and women. In 2014, 96,830 cases of colon cancer and 40,000 cases of rectal cancer are expected to be diagnosed, of which about 50,310 deaths are expected to occur. Moreover, the mortality rates for colorectal cancer have declined over the past two decades indicating waning incidence rates and improvements in early detection and treatment (American Cancer Society, 2014).

Various human colon carcinoma cell lines when treated with apigenin show cell growth inhibition and G_2/M cell cycle arrest, which was associated with inhibition of p34 (cdc2) kinase, and with reduced accumulation of p34 (cdc2) and cyclin B1 proteins (Wang et al., 2000). Apigenin has also been found to protect against the formation of aberrant crypt foci induced by azoxymethane in CF-1 mice and to reduce cancer development in this model. These results suggest that apigenin may prevent colorectal cancer through blocking of cell-cycle progression (Wang et al., 2004). In other research, Zhong et al. wanted to develop the best flavonoids for the inhibition of cell growth and found apigenin to be the most promising compound in colorectal cancer cell growth arrest. Subsequently, apigenin induced proapoptotic proteins (NAG-1 and p53) and cell cycle inhibitor (p21). The kinase pathways, including PKCδ and 'ataxia telangiectasia mutated' (ATM) play an important role in activating these proteins. Apigenin was also found to reduce polyp numbers, followed by increasing p53 activation through phosphorylation in animal models (Zhong et al., 2010). When a study was performed to analyze the effects of apigenin and underlying mechanisms on the growth of colorectal cancer xenografts in nude mice, derived from SW480, apigenin enhanced the expression of 'Fas-associated protein with death domain' (FADD) and induced its phosphorylation, which might cause apoptosis of colorectal cancer cells and inhibition of tumor growth (Wang et al., 2011).

Similarly, apigenin inhibited tumor growth and metastasis, both *in vitro* and *in vivo*, with different colorectal adenocarcinoma cell lines. In these cases, apigenin upregulated transgelin (TAGLN) and hence downregulated matrix metallopeptidase (MMP-9) expression through decreasing phosphorylation of Akt at Ser473 and, in particular Thr308, to prevent cell proliferation and migration (Chunhua et al., 2013).

When the relationship of autophagy in apigenin-induced apoptotic cell death of HCT116 human colon cancer cells was explored, apigenin showed apoptotic and autophagy-inducing effects in HCT116 colon cancer cells. Autophagy plays a cytoprotective role in apigenin-induced apoptosis, and the combination of apigenin and autophagy inhibitor may be a promising strategy for regulating colon cancer (Lee et al., 2014). A previous study on colon cancer cells with apigenin had also shown that the combined treatment with apigenin and ABT-263 inhibited tumor growth by up to 70% without noticeable adverse effects, whereas each agent alone inhibited by ~30% (Shao et al., 2013). Similarly, evaluation of the effects of apigenin on apoptotic pathway gene expression in the colon cancer cell line (HT-29) revealed that treatment with apigenin alone resulted in cell cycle arrest through activation of the caspase cascade and stimulation of apoptosis in HT-29 cells. The effect was further increased when apigenin was used along with 5-fluorouracil (Turktekin et al., 2011). Lefort and Blay investigated the possible actions of apigenin on the levels of anti-metastatic protein CD26 in human colorectal carcinoma cells. CD26 is down-regulated in colorectal carcinoma. They observed that apigenin significantly upregulated cell surface CD26 in HT-29 and HRT-18 human colorectal cancer cells and the effect was enhanced when apigenin was combined with chemotherapeutic agents, such as irinotecan, 5-fluorouracil, and oxaliplatin. These results might help in reducing tumor growth and metastasis (Lefort and Blay, 2011).

Endometrial (Uterine Corpus) Cancer

Endometrial cancer is the most common gynecological malignancy. It usually occurs in the endometrium (lining of the uterus). On the basis of clinical, endocrine, and epidemiological observations, endometrial carcinomas have been classified as type I or II (Bokhman, 1983). Endometrial carcinoma has also been classified on the basis of histopathological characteristics (Silverberg et al., 2003). Some relationships have been observed between the subtypes in these two classification systems but they are imperfect. The Cancer Genome Atlas Research Network and some other scientists have recently classified endometrial cancers into four categories: POLE (a catalytic subunit of DNA polymerase epsilon involved in nuclear DNA replication and repair) ultramutated, microsatellite instability hypermutated, copy-number low, and copy-number high (Cancer Genome Atlas Research Network, 2013). This classification was based on an integrated genomic, transcriptomic, and proteomic characterization of 373 endometrial carcinomas using array-

and sequencing-based technologies. However, there is still a need for an integrated classification that incorporates the traditional and newer genomic classifications of endometrial carcinoma, which might eventually enable advancements in treatments tailored to specific subgroups and could potentially facilitate the delivery of precision medicine to patients with endometrial cancer (Murali et al., 2014).

The American Cancer Society has estimated that 52,630 cases of endometrial cancer will be diagnosed in 2014 in the United States, with 8,590 cases succumbing to the disease. The rates of incidence and deaths have increased by 1.5% from 2006-2010 among women younger than 50 years (American Cancer Society, 2014).

O'Toole et al. identified genomic aberrations in endometrial cancer cells when treated with phytoestrogenic compounds, such as apigenin, tectorigenin, and irigenin, using array-based comparative genomic hybridization. Results showed that over 20% of the array genes were aberrant in the cells treated with β-estradiol, tectorigenin, and irigenin, compared with 3% in those treated with the same concentration of apigenin, suggesting that apigenin may play a role in the treatment of endometrial carcinoma and in the treatment of postmenopausal women (O'Toole et al., 2005).

Gastric Cancer

Gastric cancer, also known as stomach cancer, is a cancer that starts in the stomach. It should not be confused with other cancers that can occur in the abdomen, like cancer of colon/rectum, liver, pancreas, or small intestine because these cancers may have different symptoms, outlooks, and treatments. Adenocarcinoma, lymphoma, gastrointestinal stromal tumor (GIST), carcinoid tumor, and squamous cell cancers are major types of stomach cancer.

According to the American Cancer Society, an estimated 22,220 people in the United States will be diagnosed with stomach cancer and 10,990 deaths will occur from this disease in 2014 (American Cancer Society, 2014). Although the incidence of gastric cancer is decreasing in Western countries, it is still one of the most common cancer types worldwide. While infection with *Helicobacter pylori* bacteria is a major cause of gastric cancer, researchers have found several other factors that make a person more prone to stomach cancer. In recent research carried out on *H. pylori*-infected gastric adenocarcinoma MKN45 cells to investigate the anti-inflammatory effects of apigenin, it was found that apigenin successfully inhibited NF-κB activation and related inflammatory factor expression, along with an increase in mucin-2 (MUC-2) expression in MKN45 cells (Wang et al., 2013). Similarly, treatments with apigenin effectively decreased *Helicobacter pylori* colonization and *Helicobacter pylori*-induced histological changes in neutrophil and monocyte infiltration and atrophic gastritis in both atrophic gastritis and gastric cancer in Mongolian desert rats (Kuo et al., 2014).

When the growth inhibition effect of apigenin on human gastric carcinoma SGC-7901 cells was assessed, apigenin dramatically suppressed the growth and clone formation of the cells, in a dose- and time-dependent manner (Wu et al., 2005).

Leukemia

Leukemia is a cancer of the blood and bone marrow. Most often, leukemia is cancer of the white blood cells, but some leukemias start in other blood cell types. Based on the cell type and rate of growth, leukemia has been categorized into four major types: acute lymphocytic (ALL), chronic lymphocytic (CLL), acute myeloid (AML), and chronic myeloid (CML). Moreover, some oncologists classify leukemia in children as different from the other types. The ACS estimated that 52,380 new cases of leukemia will be diagnosed in 2014, and that 24,090 patients will succumb to the disease.

When different leukemic cell lines were treated with apigenin, Vargo et al. showed that apigenin provided selective activity to promote caspase-dependent apoptosis in leukemia cells and uncovered an essential role of PKCδ during the induction of apoptosis by apigenin (Vargo et al., 2006). Similarly, in an analysis of the molecular consequences of apigenin treatment in two types of leukemia, the myeloid and erythroid subtypes, apigenin proved itself as a potential chemopreventative agent by inducing leukemia cell-cycle arrest (Ruela-de-Sousa et al., 2010). However, apigenin should be examined carefully because it may also potentially interfere with cancer treatment.

In other research on apigenin-induced apoptosis in leukemia cells, the effect of apigenin in the regulation of heat-shock protein-27 (Hsp27) was studied because high expression of the small Hsp27 in leukemia contributes to the resistance of these cells to cancer treatments. Although apigenin did not affect the expression of Hsp27, it induced a bimodal phosphorylation on Ser78 and Ser82. The phosphorylation in the beginning was regulated by p38. At later times, Hsp27 phosphorylation was dependent on p38 activity and for some residues on PKCδ. Silencing of p38 expression reduced apigenin-induced phosphorylation on Ser15, Ser78, and Ser82, whereas silencing of PKCδ expression reduced the phosphorylation on Ser15 and Ser82, without affecting Ser78. Furthermore, apigenin-induced PKCδ activity was found to be mediated by p38 and the phosphorylation of Hsp27 significantly increased the susceptibility of leukemia cells to apigenin-induced apoptosis (Gonzalez-Mejia et al., 2010). Another study showed that apigenin induced apoptosis in primary effusion lymphoma cells via suppression of the Akt pathway (Hussain et al., 2010). Apigenin-induced Akt inactivation represents a primary event resulting in JNK activation, and downregulation of Mcl-1 and Bcl-2, culminating in caspase activation, and apoptosis (Budhraja et al., 2012).

Mahbub et al. recently compared the anti-proliferative and pro-apoptotic effects of apigenin and seven other polyphenols on a panel of leukemic cell lines (KG-1a, HL-60, THP-1, K562, Jurkat, CCRF-CEM, MOLT-3), one histolytic lymphoma cell line (U937), and a non-tumor hematopoietic stem progenitor cells ($CD34^+$) from cord blood (Mahbub et al., 2013). Results showed that apigenin was able to induce apoptosis in all leukemia cells, with AP_{50} values in the range of 35-130 μM in lymphoid cell lines and 84-235 μM in myeloid cell lines. Apigenin was also capable of inducing an increase in caspase-3, which is an indication of early apoptosis, and morphological evidence of late apoptosis, in all leukemia cell lines, including the KG-1a and K562 cells that were resistant to emodin, quercetin, and *cis*-stilbene. Furthermore, Jayasooriya et al. revealed that ROS were not essential for suppression of apigenin-mediated apoptosis associated with the activation of caspases and regulation of telomerase activity via suppression of human telomerase reverse transcriptase (hTERT) (Jayasooriya et al., 2012).

Liver Cancer

The liver is the largest internal organ in the human body. Its functions include processing and storing necessary nutrients absorbed by the intestines, collecting and filtering blood from the intestines, removing toxins from the body, producing blood clotting factors, chemically metabolizing some nutrients into energy or to repair and build tissue, and helping maintain the proper sugar level in the body. Thus, no one can live without a liver. The liver is unusually accessible to cancer cells travelling in the bloodstream because all the blood in the body must pass through it.

Based on the origin, liver cancer can be of two types. Cancer that begins in the liver is called primary liver cancer, whereas one that has spread to the liver from other parts of the body is known as secondary or metastatic liver cancer. Most liver cancers are secondary or metastatic. The common cancers that spread to the liver include breast cancer, lung cancer, bowel cancer, and pancreatic cancer. According to the ACS, more than 700,000 people are diagnosed with liver cancer each year globally and more than 600,000 deaths occur per year due to it. This cancer is much more common in sub-Saharan Africa and Southeast Asian countries. (American Cancer Society, 2014).

An evaluation of the anti-hepatoma properties of apigenin in three human hepatoma cell lines (Hep G2, Hep 3B and PLC/PRF/5) showed that its activity was as high as that of a commercial anti-hepatoma agent, 5-flurouracil (5-FU), and its apoptotic mechanism may be mediated through a p53-dependent pathway and involve the induction of p21 expression (Chiang et al., 2006). When the role of reactive oxygen species (ROS) as a facilitator of the apoptosis caused by apigenin was evaluated in Hep G2 human hepatoma cells alone, it was seen that ROS generated through the activation of NADPH oxidase might have played a vital role (Choi et al., 2007). The inhibitory role for apigenin in human hepatoma cells, associated with induction of G_2/M arrest and apoptosis, resulted from alterations in gene expression patterns and especially downregulation of two tumorigenesis-related genes, IL-4R and USP18 (Cai et al., 2011). Other researchers have shown that IL-4R are overexpressed on the surface of different human solid tumors, including lung, ovarian, glioblastoma, and head and neck squamous cell carcinomas (Kioi et al., 2005; Melstrom et al., 2008). Inhibition of IL-4R signaling has been proposed as a novel treatment for some cancers (Kioi et al., 2005; Todaro et al., 2008). However, Singh et al. demonstrated that apigenin induced oxidative stress through a different pathway, resulting in liver toxicity (Singh et al., 2012).

Lung Cancer

Cancer that forms in tissues of the lung, usually in the cells lining the air passages, is called lung cancer. The major types of lung cancer are non-small cell lung cancer and small-cell lung cancer. About 85% of lung cancers are non-small cell lung cancers. Small cell lung cancer, also called oat cell cancer, accounts for ~10-15% of lung cancers. Besides the two major types, lung carcinoid tumor is the third type, accounting for ~5% of lung cancers. Most of these tumors grow slowly and rarely spread.

Lung cancer is responsible for more deaths than any other cancer in both men and women. Every year, 1.37 million people die from this disease worldwide. An estimated 159,260 Americans are expected to die from lung cancer in 2014. Smoking is the cause of most lung cancers, but there are other risk factors and causes that can increase the chances of developing the disease.

Similar to many other cancers, the role of reactive oxygen species, Ca^{2+}, caspases, Bax proteins, and mitochondria membrane potential were investigated in apigenin-induced apoptosis in human lung adenocarcinoma A549 cells (Lu et al., 2010). The results showed that apigenin induced morphological changes, decreased the percentage of viable cells, and induced apoptosis in a dose- and time-dependent manner, along with increases in the levels of caspases-3, -8 and, -9, involved in apigenin-induced apoptosis. Furthermore, apigenin promoted cytochrome *c* levels and induced dysfunction of mitochondria, leading to the release of cytochrome *c*, AIF and Endo G, causing activation of caspases-9 and -3, and then apoptosis in A549 cells. Similarly, the roles of apigenin and of the leptin/leptin receptor pathway on proliferation and on apoptosis in A549 cells were studied by Bruno et al. (Bruno et al., 2011). This study demonstrated that A549 cells expressed the leptin/leptin receptor pathway and its expression was upregulated by apigenin. Furthermore, apigenin alone or combined with leptin decreased cell proliferation significantly and increased the spontaneous release of ROS, with augmented cell death and apoptosis, the latter even in the presence of lung adenocarcinoma pleural fluid. Das et al. in their research on apigenin-induced apoptosis in human malignant melanoma cell line (A375) and A549 cells showed that apigenin did not directly target the mitochondrial oxidative phosphorylation system but rather acted at an upstream step to activate the mitochondrial apoptotic pathway in A549 cells (Das et al., 2012).

Ren and Tang demonstrated that apigenin inhibited cell proliferation in various types of lung cancer cell lines in a dose-dependent manner (Ren and Tang, 2011). They further recommended combination treatment with apigenin and anti-tumor drugs for human lung cancer cells as an effective approach.

Neuroblastoma

Neuroblastoma is a solid cancerous tumor that begins in early nerve cells (neuroblasts) of the sympathetic nervous system in infants and young children. It is the third most common childhood cancer and represents 7% of total cancer cases in the age group 0-14, although it becomes very uncommon after age 10. Neuroblastoma may be caused by genetics. Although epidemiological studies have investigated environmental factors that may be associated with neuroblastoma, no strong or consistent risk factor has yet been identified.

The ability of apigenin to inhibit growth and induce apoptosis in the human neuroblastoma cell lines NUB-7, LAN-5, and SK-N-BE(2) was investigated by Torkin et al. Apigenin inhibited colony forming ability and survival, and induced apoptosis in NUB-7 and LAN-5 cells. However, apigenin did not inhibit the survival of primary sympathetic neurons, indicating that it was not toxic to non-transformed cells. Torkin et al. proposed that the mechanism of action involved p53, because it increased the levels of p53 and the p53-induced gene products p21WAF1/CIP1 and Bax. Moreover, apigenin induced cell death and apoptosis in neuroblastoma cells expressing wild-type but not mutant p53. Apigenin also increased caspase-3 activity and PARP cleavage, and Z-VAD-FMK, a broad-spectrum caspase-3 inhibitor, rescued NUB-7 cells from apigenin-mediated apoptosis, indicating that apigenin induced apoptosis in a caspase-dependent manner. Furthermore, overexpression of Bcl-X(L) rescued NUB-7 from apigenin-induced cell death, suggesting that Bax activity is important for the action of apigenin (Torkin et al., 2005).

In research by Karmakar et al., the efficacy of the small molecule Bcl-2 inhibitor HA14-1 (HA) and the dietary flavonoid apigenin in human malignant neuroblastoma cells was tested (Karmakar et al., 2009). They demonstrated that the combination of HA and apigenin could be used for regulating angiogenic factors and activation of the extrinsic and intrinsic pathways of apoptosis in malignant neuroblastoma cells. Similarly, Mohan et al. explored the efficacy of the synthetic retinoid N-(4-hydroxyphenyl) retinamide (4-HPR) and apigenin in serum-starved human malignant neuroblastoma cells and showed that the combination of 4-HPR and apigenin worked synergistically to suppress autophagy and promoted apoptosis in human malignant neuroblastoma cells (Mohan et al., 2011). Furthermore, the same group studied the effects of Krüpple-like factor 4 (KLF4) overexpression and apigenin treatment in human malignant neuroblastoma SK-N-DZ and IMR-32 cell lines. The results showed that 100 nM KLF4 plasmid and 25 μM apigenin synergistically inhibited the growth of SK-N-DZ and IMR-32 cells and that treatment with a combination of KLF-4 and apigenin significantly increased the amounts of apoptosis in both cell lines when compared with the control vector or single treatment. The combination therapy also decreased expression of the anti-apoptotic proteins Bcl-2 and Mcl-1, increased expression of the pro-apoptotic proteins Bax, Noxa, and Puma, upregulated p53, and caused activation of caspase-3 for cleavage of the inhibitor of caspase-activated DNase (ICAD) leading to completion of the apoptosis machinery. Moreover, the combination of KLF4 overexpression and apigenin treatment was highly effective in inhibiting migration of both neuroblastoma cell lines and was associated with downregulation of matrix metalloproteinases (MMPs), such as MMP-2 and MMP-9 (Mohan et al., 2013). Combination therapy of a similar type was also applied to malignant neuroblastoma cells having N-Myc amplification. First, the expression of the N-Myc oncogene was decreased using a short hairpin RNA (shRNA) plasmid, which would help to increase the anti-tumor efficacy of apigenin in malignant neuroblastoma SK-N-DZ and SK-N-BE-2 cell lines harboring N-Myc amplification. Then, the combination of N-Myc knockdown and apigenin most effectively induced morphological and biochemical features of apoptotic death, including the prevention of cell migration and decreased N-Myc-driven survival, angiogenic, and invasive factors (Hossain et al., 2013).

Ovarian Cancer

Ovarian cancer begins when normal cells in an ovary change and grow uncontrollably, forming a mass called a tumor. There are three types of ovarian tissue that can produce cancers: epithelial cells, which cover the ovary, stromal cells, which produce hormones, and germ cells, which become ova. About 85-90% of ovarian cancers are epithelial carcinomas (Hanna et al., 2003). Removing the ovary or the part of ovary where tumor is located can treat a noncancerous (benign) ovarian tumor. An ovarian cyst, which forms on the surface of the ovary, is different from a non-cancerous tumor and usually resolves without treatment.

According to the Ovarian Cancer 2014 Report, this is the seventh most common cancer in women worldwide. Approximately 239,000 cases were recorded in 2012, which is ~4% of all new cases of cancer in women. This cancer is usually fatal, and is the 8th most common cause of cancer death in women worldwide (14th overall) (Ferlay et al., 2013). With age, the risk for the disease increases, but slows down after menopause. Only 10-15% of cases occur before menopause whereas the incidence of germ cell cancers, which are uncommon, is higher in women aged between 15 and 35 (Hanna et al., 2003). Nevertheless, ovarian cancer remains the most lethal among the gynecological cancers due to lack of early detection methods and effective treatments for late-stage cancers (Jemal et al., 2005).

Many studies have been carried out to evaluate the relationship between dietary flavonoid intake and ovarian cancer risk (McCann et al., 2003; Zhang et al., 2004; Chang et al., 2007). In a study carried out to analyze the association between intake of five common dietary flavonoids (myricetin, kaempferol, quercetin, luteolin, and apigenin) and incidence of epithelial ovarian cancer among 66,940 women, the results suggested that dietary intake of certain flavonoids might reduce ovarian cancer risk (Gates et al., 2007). Hu et al. demonstrated that apigenin inhibited the expression of focal adhesion kinase (FAK) in ovarian cancer A2780 cells, which lead to inhibition of *in vitro* migration and invasion and *in vivo* metastasis of A2780 cells (Hu et al., 2008). Furthermore, the inhibition of proliferation and tumorigenesis of A2780 cells by apigenin was via suppression of expression of 'inhibitor of differentiation or DNA binding protein 1' (Id1) through 'activating transcription factor 3' (ATF3) (Li et al., 2009). Similarly, a study on the effects of 12 flavonoids including apigenin and other substances on cell proliferation and vascular endothelial growth factor (VEGF) expression in human ovarian cancer cells revealed that these flavonoids inhibited ovarian cancer cell proliferation and VEGF expression at both the mRNA and protein levels; the regulatory mechanism occurred mainly at the mRNA level (Luo et al., 2008).

He et al. developed an orthotopic ovarian tumor model in nude mice to test the effect of apigenin oral administration, and found that oral administration of apigenin inhibited ovarian tumor micrometastasis in liver, lung, small intestine, and stomach to differing degrees. While analyzing the mechanism, it was found that apigenin inhibited MMP-9 expression significantly and p-AKT and p-p70S6K1 levels in the tumor tissues compared with the control group. Furthermore, the downregulation of MMP-9 by apigenin was mediated by the AKT/p70S6K1 pathway (He et al., 2012).

Pancreatic Cancer

The pancreas is an organ located behind the stomach. It is ~6 inches long and 2 inches wide. It contains two different types of gland: exocrine and endocrine. The exocrine gland, which is made up of ducts and acini, produces pancreatic juice. The endocrine gland is made up of specialized cells called islets of Langerhans, which make important hormones, such as insulin and glucagon.

There are various types of pancreatic cancer, depending on whether the cancer begins in the exocrine or endocrine gland. Exocrine tumors are the most common type of pancreatic cancer whereas tumors of the endocrine glands are rare. Pancreatic cancer is the seventh most common cause of cancer death worldwide, with more than 330,000 deaths from pancreatic cancer in 2012, which is 4% of all deaths from cancer.

Many chemotherapeutic agents have been used to treat pancreatic cancer without success. Apigenin caused both time- and dose-dependent inhibition of DNA synthesis and cell proliferation in four pancreatic cancer cell lines through suppression of cyclin B-associated cdc2 activity and G2/M arrest (Ujiki et al., 2006). The same group also investigated the role of apigenin on geminin (a novel protein that plays a critical role in preventing abnormal DNA replication by binding to and inhibiting the essential replication factor Cdt1) expression and other proteins involved in replication (Cdt6, Cdt1, and MCM7) in the pancreatic cancer cell lines CD18 and S2013 (Salabat et al., 2008). The results showed that geminin was overexpressed in pancreatic cancer compared with normal adjacent pancreatic tissue and this expression was downregulated *in vitro* by apigenin. Furthermore, they demonstrated that apigenin decreased glucose uptake and downregulated the glucose transporter 1 (GLUT-1) in human pancreatic cancer (Melstrom et al., 2008). Melstrom et al. further showed that apigenin inhibited HIF-1α, GLUT-1, and VEGF mRNA and protein expression in pancreatic cells in both normoxic and hypoxic conditions (Melstrom et al., 2011).

Additionally, apigenin in combination with gemcitabine inhibited pancreatic cancer cell growth via cell cycle arrest, downregulation of prosurvival factor pAkt, and induction of apoptosis (Strouch et al., 2009). Earlier, in 2008, Lee et al. also evaluated the anti-tumor efficacy of combination therapy with gemcitabine and apigenin in pancreatic cancer cell lines (MiaPaca-2 and AsPC-1) (Lee et al., 2008). The results of *in vitro* and *in vivo* studies showed that the combination of gemcitabine and apigenin enhanced anti-tumor efficacy through Akt and NF-κB activity suppression and apoptosis induction. King et al. examined the anti-neoplastic mechanisms of apigenin with a particular focus on p53 function (King et al., 2012). They reported that despite mutated p53 in pancreatic cancer cells (BxPC-3 and MiaPaCa-2), p53 retained some of its function, which was augmented following treatment with apigenin and further suggested that cell cycle arrest and apoptosis induction might be mediated by transcription-independent p53 function via interactions with BclXL and PUMA. Similarly, flavonoids, limonoids, phenolic acids, and ascorbic acid were tested for cytotoxic effects on human pancreatic cancer cells (BxPC-3 and PANC-1) (Johnson and de Mejia, 2013) in which apigenin stood out as the most potent agent. Apigenin induced pancreatic cell death through inhibition of the glycogen synthase kinase-3β/NF-κB signaling pathway, arrested cell cycle at G2/M phase with concomitant decrease in the expression of cyclin B1, activated the mitochondrial pathway of apoptosis, and modified the expression of apoptotic proteins. Furthermore, apigenin highly upregulated the expression of cytokine genes IL17F, LTA, IL17C, IL17A, and IFNB1 in BxPC-3 cells, which possibly contributed to the anti-cancer properties. Recently, Wu et al. also demonstrated that apigenin had the potential to inhibit inhibitor of κB kinase (IKK)-β-mediated NF-κB activation (Wu, et al., 2014).

Prostate Cancer

Prostate cancer is a malignant tumor that begins in the prostate gland, a walnut-sized gland located behind the base of a man's penis, in front of the rectum, and below the bladder. Some prostate cancers grow very slowly and may not cause problems for years. Prostate cancer is somewhat unusual, compared with other types of cancer, because many tumors do not spread from the prostate. Moreover, often even metastatic prostate cancer can be treated successfully, allowing patients to live with good health for several years. More than 1.11 million men were estimated to have been diagnosed with prostate cancer in 2012. The strongest risk factor for prostate cancer is age, with very low risk in men under the age of 50, but risk increases with age thereafter.

Shukla and Gupta showed that apigenin-induced cell cycle arrest in human prostate cancer cells was mediated by modulation of MAPK, PI3K-Akt, and loss of cyclin D1, D2, and E, and their regulatory partners CDK 2, 4, and 6, operative in the G0-G1 phase of the cell cycle (Shukla and Gupta, 2007). Furthermore, the effects of apigenin on prostate carcinogenesis in "transgenic adenocarcinoma of the mouse prostate" (TRAMP) were investigated by the same group, showing effective suppression of prostate carcinogenesis by apigenin in TRAMP mice by blocking β-catenin signaling (Shukla et al., 2007). In recent publications by the same group, they have presented the evidence that apigenin effectively suppressed prostate cancer progression by targeting the PI3K/Akt/FoxO-signaling pathway (Shukla et al., 2014). Similarly, their results suggested that apigenin targeted the inhibitor of apoptosis proteins and Ku70-Bax interaction in the induction of apoptosis in prostate cancer cells and in an athymic nude mouse xenograft model demonstrating *in vivo* efficacy (Shukla et al., 2014).

In an investigation to study the effects of apigenin on expression of hypoxia-inducible factor-1 (HIF-1) in human metastatic prostate PC3-M cancer cells, downregulation of HIF-1α protein levels in human metastatic prostate PC3-M cells by apigenin was confirmed, which was linked with destabilization of the HIF-1α protein, a decrease in HIF-1α mRNA levels, and prevention of HIF-1 target gene activation. Moreover, apigenin affected the activation of Akt and inhibition of GSK-3β (Mirzoeva et al., 2008). Similar results were observed by Kaur et al. when they investigated the mechanisms of apigenin action on human prostate cancer PC-3 cells (Kaur et al., 2008). They showed that treatment of PC-3 cancer cells with apigenin resulted in dose- and time-dependent decreases in Akt phosphorylation and induction of apoptosis in PC-3 cells. Furthermore, Shukla and Gupta (Shukla and Gupta, 2009) showed that inhibition of cell proliferation and induction of apoptosis by apigenin were mediated, at least in part, by its ability to inhibit insulin-like growth factor (IGF)-I/IGF-IR signaling and the PI3K/Akt pathway. Franzen et al. evaluated the effect of apigenin on motility, invasion, and its mechanism of action in metastatic prostate carcinoma cells (PC3-M) (Franzen et al., 2009). The results revealed that apigenin inhibited motility and invasion of PC3-M cells, disrupted actin cytoskeleton organization, and inhibited FAK/Src signaling. Recently, Zhu et al. reported that apigenin suppressed the proliferation and inhibited migration and invasive potential of the DU145 prostate cancer cells in a dose- and time-dependent manner, which was associated with the epithelial-mesenchymal transition (Zhu et al., 2014). Even the co-administration of vicenin-2 (apigenin-6,8-di-C-glucoside) and docetaxel was found more effective than either of the agents alone in androgen-independent prostate cancer (Nagaprashantha et al., 2011). Additionally, apigenin was found to inhibit class I histone deacetylase (HDAC) expression, mainly HDAC1 and HDAC3, causing increased acetylation of histone H3 and H4, improving accessibility of the promoter region of the p21/waf1 gene causing cell cycle arrest in prostate cancer cells, and inhibiting tumor growth of PC-3 xenografts in nude mice (Pandey et al., 2012).

Skin Cancer

The skin is the body's largest organ, which protects against infection and injury and helps to regulate body temperature. There are three main types of skin cancer: basal cell carcinoma, squamous cell carcinoma, and melanoma. The first two types of skin cancers are highly curable, but can be disfiguring and costly, whereas the last one is the most common, most dangerous, and causes the most deaths. The incidence of skin cancers has been increasing at an alarming rate over past decades, likely because the rate of ozone layer depletion is also increasing. It is estimated that every year, about 2-3 million non-melanoma skin cancers and 132,000 melanoma skin cancers occur globally.

Wei et al. investigated the effect of the topical application of apigenin on skin tumorigenesis and established apigenin as a potent inhibitor of epidermal ornithine decarboxylase and skin tumor promotion in mice (Wei et al., 1990). Apigenin also acted as an effective inhibitor of chemically and ultraviolet B (UVB) light-induced skin tumorigenesis in mice and human when applied topically (Birt et al., 1997; Abu-Yousif, 2008; Byun et al., 2013). Das et al. explored the possibility of enhanced anti-carcinogenic effect of apigenin against UVB- and benzo(a)pyrene-induced skin tumors and mitochondrial dysfunction in mice (Das et al., 2013). They uploaded apigenin with poly(lactic-co-glycolide) (PLGA) nanoparticles for this purpose and found that apigenin on the nanoparticles showed 'better' effects than apigenin alone, due to their smaller size and faster mobility. Additionally, apigenin uploaded on nanoparticles reduced tissue damage and the frequency of chromosomal aberrations, and increased ROS accumulation to mediate mitochondrial-apoptosis through modulation of several apoptotic markers and mitochondrial matrix swelling.

Exposure of mouse keratinocytes to apigenin induced G2/M cell-cycle arrest and accumulation of the p53 tumor suppressor protein with increased expression of p21/WAF1. This arrest was complemented by inhibition of p34 (cdk2) kinase protein levels and activity, which was found to be independent of p21/WAF1 (McVean et al., 2000). In human diploid fibroblasts, apigenin produced G1 cell-cycle arrest by inhibiting cdk2 kinase activity and inducing p21/WAF1. Li et al. established a short-term *in vivo* system to evaluate topical formulations of apigenin and to determine whether apigenin was effective when delivered as a topical preparation to local skin lesions (Li et al., 1996). Furthermore, Tong et al. reported activation of AMPK through CaMKKβ by apigenin in human keratinocytes, which resulted in inhibition of mammalian target of rapamycin (mTOR) signaling and induction of autophagy in keratinocytes (Tong et al., 2012). The results showed that topical application of apigenin was capable of targeting local tissue. Moreover, Chan et al. showed that apigenin induced SCC25 cell apoptosis via the upregulation of both tumor necrosis factor receptor (TNF-R) and TNF-related apoptosis-inducing ligand receptor (TRAIL-R) signaling pathways, downregulation of Bcl-2, activation of caspase-3, and had a synergistic effect on the inhibition of cell proliferation in combination with 5-fluorouracil or cisplatin on human head and neck squamous cell carcinoma (Chan et al., 2012). Caltagirone et al. also reported that apigenin co-operated positively with cisplatin in inhibiting melanoma growth *in vivo* (Caltagirone et al., 2000). Recently, Paredes-Gonzalez et al. evaluated the epigenetic effect of apigenin in skin cancer chemoprevention (Paredes-Gonzalez et al., 2014). They used bisulfite genomic DNA sequencing and methylated DNA immunoprecipitation to explore the demethylation effect of apigenin at 15 CpG sites in the Nrf2 promoter in mouse skin epidermal JB6 P^+ cells. The results from their experiments revealed that apigenin can restore the silenced status of Nrf2 in skin epidermal JB6 P^+ cells by CpG demethylation, coupled with attenuated DNA methyltransferase and histone deacetylase activity. All these results may offer new therapeutic insights into the prevention of skin cancer by the dietary flavonoid apigenin.

Thyroid Cancer

The thyroid gland is part of the endocrine system, which regulates hormones in the body. It absorbs iodine from the bloodstream to produce thyroid hormones, which regulate a person's metabolism. The thyroid gland comprises two types of cells: follicular cells and C cells; the former are responsible for the production of thyroid hormone whereas the latter produce calcitonin, a hormone that contributes to calcium metabolism. There are four different types of thyroid cancer: papillary thyroid cancer, follicular thyroid cancer, medullary thyroid cancer and anaplastic thyroid cancer. The cases of thyroid cancer have been increasing worldwide over the past few decades.

The effects of some selected flavonoids on human thyroid carcinoma cell lines, UCLA NPA-87-1 (papillary carcinoma), UCLA RO-82-W-1 (follicular carcinoma), and UCLA RO-81A-1 (anaplastic carcinoma) were investigated by Yin et al. (Yin et al., 1999). Their results showed that apigenin and luteolin were the most potent inhibitors of cell proliferation. They assumed the inhibitory activity of certain flavonoid compounds was mediated via the anti-estrogen binding site (AEBS) and/or type II estrogen binding site (EBS). However, effective inhibition of UCLA RO-81A-1 cells, which lack both the AEBS and estrogen receptor, by apigenin and luteolin, resulted in a hypothesis that other mechanisms were also involved. Similarly, the same group confirmed that the inhibitory effect of apigenin on UCLA-RO-81A-1 cell proliferation was associated with inhibition of both EGFR tyrosine autophosphorylation and phosphorylation of its downstream effector, mitogen activated protein kinase (Yin et al., 1999).

Schröder-van der Elst et al. assessed the role of flavonoids on iodide transport and the growth of the human follicular thyroid cancer cell line (FTC133), which was stably transfected with the human Na^+/I^- symporter (hNIS) and found that apigenin, luteolin, and kaempferol decreased NIS mRNA expression (Schröder-van der Elst et al., 2004). When the effect of apigenin on cell survival and c-Myc expression in FRO anaplastic thyroid carcinoma (ATC) cells was analyzed, the results showed that apigenin-induced apoptosis facilitated via c-Myc with simultaneous phosphorylation of p53 and p38 in FRO ATC cells (Kim et al., 2013). Recent research by Lakshmanan et al. showed that apigenin enhanced the iodide influx rate increased by Akt inhibition in thyroid cells under acute thyrotropin (TSH) stimulation. The increase in radioiodide accumulation by apigenin with Akt inhibition was also observed in thyroid cells expressing BRAF (V600E) and in primary cultured thyroid tumor cells from TRβ (PV/PV) mice (Lakshmanan et al., 2014).

CONCLUSION

Various preclinical findings and results of several *in vitro* and *in vivo* studies have suggested the potential role of apigenin in the prevention and treatment of many types of cancer. Many mechanisms of action have been identified for apigenin-mediated cancer prevention and therapy, which include estrogenic/anti-estrogenic activity, anti-proliferative activity, prevention of oxidation, induction of detoxification enzymes, induction of cell-cycle arrest, regulation of the host immune system, and changes in cellular signaling (Patel et al., 2007; Amin et al., 2009). Beneficial effects have been described also by combining apigenin with either standard chemotherapeutic drugs or flavonoids, leading to reduction in dosage and concomitant toxicity while targeting specific mechanisms. This further helps in minimizing the genotoxic damage caused by 'standard' chemotherapeutics to normal cells, which ultimately reduces the chance of developing secondary cancers (Lefort and Blay, 2013; Strouch et al., 2009). However, some discrepancies have been observed in the results, likely due mostly to the use of different protocols and experimental conditions in different laboratories. Nevertheless, apigenin is one of the most bioactive plant flavones and widely distributed in common fruits, vegetables, and beverages; its consumption through the diet is highly recommended. Importantly, a number of research findings suggest that apigenin has potential for development as a promising cancer chemopreventive agent in near future.

References

Abu-Yousif, AO; Smith, KA; Getsios, S; Green, KJ; Van Dross, RT; Pelling, JC. Enhancement of UVB-induced apoptosis by apigenin in human keratinocytes and organotypic keratinocyte cultures. *Cancer Research,* 2008, 68, 3057-3065.

American Cancer Society. Cancer Facts and Figures 2014. Atlanta: American Cancer Society; 2014.

Amin, AR; Kucuk, O; Khuri, FR; Shin, DM. Perspectives for cancer prevention with natural compounds. *Journal of Clincal Oncology,* 2009, 27, 2712-2725.

Are, C; Rajaram, S; Are, M; Raj, H; Anderson, BO; Chaluvarya Swamy, R; Vijayakumar, M; Song, T; Pandey, M; Edney, JA; Cazap, EL. A review of global cancer burden:trends, challenges, strategies, and a role for surgeons. *Journal of Surgical Oncology,* 2013, 107, 221–226.

Aune, D; Chan, DS; Vieira, AR; Rosenblatt, DA; Vieira, R; Greenwood, DC; Norat, T. Fruits, vegetables and breast cancer risk: a systematic review and meta-analysis of prospective studies. *Breast Cancer Research and Treatment,* 2012, 134, 479–493.

Aune, D; Lau, R; Chan, DS; Vieira, R; Greenwood, DC; Kampman, E; Norat, T. Nonlinear reduction in risk for colorectal cancer by fruit and vegetable intake based on meta-analysis of prospective studies. *Gastroenterology,* 2011,141, 106–118.

Berghőfer, R; Hőlzl, J. Biflavonoids in *Hypericum perforatum*; Part 1. Isolation of I3, II8-Biapigenin, *Planta Medica,* 1987, 53, 216-217.

Bevilacqua, L; Buiarelli; F; Coccioli, F; Jasionowska, R. Identification of compounds in wine by HPLC-tandem mass spectrometry. *Annali di Chimica,* 2004, 94, 679-689.

Birt, DF; Hendrich, S; Wang, W. Dietary agents in cancer prevention: flavonoids and isoflavonoids. *Pharmacology and Therapeutics,* 2001, 90, 157-177.

Birt, DF; Mitchell, D; Gold, B; Pour, P; Pinch, H. Inhibition of ultra violet light induced skin carcinogenesis in SKH-1 mice by apigenin, a plant flavonoid. *Anticancer Research,* 1997, 17, 85-91.

Bokhman, JV. Two pathogenetic types of endometrial carcinoma. *Gynecologic Oncology,* 1983, 15, 10-17.

Bonfili, L; Cecarini, V; Amici, M; Cuccioloni, M; Angeletti, M; Keller, JN; Eleuteri, AM. Natural polyphenols as proteasome modulators and their role as anti-cancer compounds. *The FEBS Journal,* 2008, 275, 5512-5526.

Bruno, A; Siena, L; Gerbino, S; Ferraro, M; Chanez, P; Giammanco, M; Gjomarkaj, M; Pace, E. Apigenin affects leptin/leptin receptor pathway and induces cell apoptosis in lung adenocarcinoma cell line. *European Journal of Cancer,* 2011, 47, 2042-2051.

Budavari, S. ed. (1997) The Merck Index (13th Ed., pp. 123-124) Whitehouse Station, NJ: *Merck & Co. Inc.*

Budhraja, A; Gao, N; Zhang, Z; Son, YO; Cheng, S; Wang, X; Ding, S; Hitron, A; Chen, G; Luo, J; Shi, X. Apigenin induces apoptosis in human leukemia cells and exhibits anti-leukemic activity in vivo. *Molecular Cancer Therapeutics,* 2012, 11, 132-142.

Butler, LM; Wu, AH; Wang, R; Koh, WP; Yuan, JM; Yu, MC. A vegetable fruit-soy dietary pattern protects against breast cancer among postmenopausal Singapore Chinese women. *The American Journal of Clinical Nutrition,* 2010, 91, 1013–1019.

Byun, S; Park, J; Lee, E; Lim, S; Yu, JG; Lee, SJ; Chen, H; Dong, Z; Lee, KW; Lee, HJ. Src kinase is a direct target of apigenin against UVB-induced skin inflammation. *Carcinogenesis,* 2013, 34, 397-405.

Cai, J; Zhao, XL; Liu, AW; Nian, H; Zhang, SH. Apigenin inhibits hepatoma cell growth through alteration of gene expression patterns. *Phytomedicine,* 2011, 18, 366-373.

Caltagirone, S; Rossi, C; Poggi, A; Ranelletti, FO; Natali, PG; Brunetti, M; Aiello, FB; Piantelli, M. Flavonoids apigenin and quercetin inhibit melanoma growth and metastatic potential. *International Journal of Cancer,* 2000, 87, 595-600.

Cancer Genome Atlas Research Network; Kandoth, C; Schultz, N; Cherniack, AD; Akbani, R; Liu, Y; Shen, H; Robertson, AG; Pashtan, I; Shen, R; Benz, CC; Yau, C; Laird, PW; Ding, L; Zhang, W; Mills, GB; Kucherlapati, R; Mardis, ER; Levine, DA. Integrated genomic characterization of endometria carcinoma. *Nature,* 2013, 497, 67-73.

Chan, LP; Chou, TH; Ding, HY; Chen, PR; Chiang, FY; Kuo, PL; Liang, CH. Apigenin induces apoptosis via tumor necrosis factor receptor- and Bcl-2-mediated pathway and enhances susceptibility of head and neck squamous cell carcinoma to 5-fluorouracil and cisplatin. *Biochimica et Biophysica Acta,* 2012, 1820, 1081-1091.

Chang, ET; Lee, VS; Canchola, AJ; Clarke, CA; Purdie, DM; Reynolds, P; Anton-Culver, H; Bernstein, L; Deapen, D; Peel, D; Pinder, R; Ross, RK; Stram, DO; West, DW, Wright, W; Ziogas, A; Horn-Ross, PL. Diet and risk of ovarian cancer in the California Teachers Study cohort. *American Journal of Epidemiology,* 2007, 165, 802–813.

Chen, D; Landis-Piwowar, KR; Chen, MS; Dou, QP. Inhibition of proteasome activity by the dietaryflavonoid apigenin is associated with growth inhibition in cultured breast cancer cells and xenografts. *Breast Cancer Research,* 2007, 9, R80.

Chiang, LC; Ng, LT; Lin, IC; Kuo, PL; Lin, CC. Anti-proliferative effect of apigenin and its apoptotic induction in human Hep G2 cells. *Cancer Letters,* 2006, 237, 207-214.

Choi, EJ; Kim, GH. Apigenin causes G_2/M arrest associated with the modulation of $p21^{Cip1}$ and Cdc2 and activates p53-dependent apoptosis pathway in human breast cancer SK-BR-3 cells. *Journal of Nutritional Biochemistry,* 2009, 20, 285–290.

Choi, SI; Jeong, CS; Cho, SY; Lee, YS. Mechanism of apoptosis induced by apigenin in HepG2 human hepatoma cells: involvement of reactive oxygen species generated by NADPH oxidase. *Archives of Pharmacal Research,* 2007, 30, 1328-1335.

Chung, FL; Schwartz, J; Herzog, CR; Yang, YM. Tea and cancer prevention: studies in animals and humans. *Journal of Nutrition,* 2003, 133, 3268S-3274S.

Chunhua, L; Donglan, L; Xiuqiong, F; Lihua, Z; Qin, F; Yawei, L; Liang, Z; Ge, W; Linlin, J; Ping, Z; Kun, L; Xuegang, S. Apigenin up-regulates transgelin and inhibits invasion and migration of colorectal cancer through decreased phosphorylation of AKT. *The Journal of Nutritional Biochemistry,* 2013, 24, 1766-1775.

Clere, N; Faure, S; Martinez, MC; Andriantsitohaina, R. Anticancer properties of flavonoids: roles in various stages of carcinogenesis. *Cardiovascular and Hematological Agents in Medicinal Chemistry,* 2011, 9, 62-77.

Czyz, J; Madeja, Z; Irmer, U; Korohoda, W; Hülser, DF. Flavonoid apigenin inhibits motility and invasiveness of carcinoma cells in vitro. *International Journal of Cancer,* 2005, 114, 12-18.

da Silveira, CV; Trevisan, MT; Rios, JB; Erben, G; Haubner, R; Pfundstein, B; Owen, RW. Secondary plant substances in various extracts of the leaves, fruits, stem and bark of *Caraipa densifolia* Mart. *Food and Chemical Toxicology,* 2010, 48, 1597–1606.

Das, S; Das, J; Samadder, A; Boujedaini, N; Khuda-Bukhsh, AR. Apigenin-induced apoptosis in A375 and A549 cells through selective action and dysfunction of mitochondria. *Experimental Biology and Medicine,* 2012, 237, 1433-1448.

Das, S; Das, J; Samadder, A; Paul, A; Khuda-Bukhsh, AR. Efficacy of PLGA-loaded apigenin nanoparticles in Benzo(a)pyrene and ultraviolet-B induced skin cancer of mice: mitochondria mediated apoptotic signaling cascades. *Food and Chemical Toxicology,* 2013, 62, 670-680.

Ferlay, J; Soerjomataram, I; Ervik, M; Dikshit, R; Eser, S; Mathers, C; Rebelo, M; Parkin, DM; Forman, D; Bray, F. GLOBOCAN 2012 v1.0, Cancer Incidence and Mortality Worldwide. 2013 (2014.11.03). Available from: http://globocan.iarc.fr.

Franzen, CA; Amargo, E; Todorović, V; Desai, BV; Huda, S; Mirzoeva, S; Chiu, K; Grzybowski, BA; Chew, TL; Green, KJ; Pelling, JC. The chemopreventive bioflavonoid apigenin inhibits prostate cancer cell motility through the focal adhesion kinase/Src signaling mechanism. *Cancer Prevention Research,* 2009, 2, 830-841.

García-Lafuente, A; Guillamón, E; Villares, A; Rostagno, MA; Martínez JA. Flavonoids as anti-inflammatory agents: implications in cancer and cardiovascular disease. *Inflammation Research,* 2009, 58, 537-532.

Gates, MA; Vitonis, AF; Tworoger, SS; Rosner, B; Titus-Ernstoff, L; Hankinson, SE; Cramer, DW. Flavonoid intake and ovarian cancer risk in a population-based case-control study. *International Journal of Cancer,* 2009, 124, 1918-1925.

Gerhäuser, C. Beer constituents as potential cancer chemopreventive agents. *European Journal of Cancer,* 2005, 41, 1941-1954.

Gonzalez-Mejia, ME; Voss, OH; Murnan, EJ; Doseff, AI. Apigenin-induced apoptosis of leukemia cells is mediated by a bimodal and differentially regulated residue-specific phosphorylation of heat-shock protein-27. *Cell Death and Disease,* 2010, 1, e64; doi:10.1038/cddis.2010.41

Gurung, RB; Kim, EH; Oh, TJ; Sohng, JK. Enzymatic synthesis of apigenin glucosides by glucosyltransferase (YjiC) from *Bacillus licheniformis* DSM 13. *Molecules and Cells,* 2013, 36, 355-361.

Hanna, N; Timmerman, R; Foster, RS; Roth, BJ; Einhorn, LH; Nichols, CR. Testis cancer. In: Kufe DW, Pollock RE, Weichselbaum RR, Bast RC, Gansler TS, Holland JF, Frei E, eds. Cancer Medicine. 6th Ed. Hamilton, Ontario: *BC Decker Inc.;* 2003:1747-1768.

Harborne, JB; Williams, CA. Advances in flavonoid research since 1992. *Phytochemistry,* 55, 481-504.

Harrison, ME; Power Coombs, MR; Delaney, LM; Hoskin, DW. Exposure of breast cancer cells to a subcytotoxic dose of apigenin causes growth inhibition, oxidative stress, and hypophosphorylation of Akt. *Experimental and Molecular Pathology*, 2014, 97, 211-217.

He, H; Chen, L; Zhai, M; Chen, JZ. Genistein down-regulates the constitutive activation of nuclear factor-kappaB in human multiple myeloma cells, leading to suppression of proliferation and induction of apoptosis. *Phytotherapy Research,* 2009, 23, 868-873.

He, J; Xu, Q; Wang, M; Li, C; Qian, X; Shi, Z; Liu, LZ; Jiang, BH. Oral administration of apigenin inhibits metastasis through AKT/P70S6K1/MMP-9 pathway in orthotopic ovarian tumor model. *International Journal of Molecular Sciences,* 2012, 13, 7271-7282.

Hertog, MG; Feskens, EJ; Hollman, PC; Katan, MB; Kromhout, D. Dietary flavonoids and cancer risk in the Zutphen Elderly Study. *Nutrition and Cancer,* 1994, 22, 175-184.

Hollman, PC; Katan, MB. Health effects and bioavailability of dietary flavonols. *Free Radical Research* 1999, 31, S75-S80.

Hossain, MM; Banik, NL; Ray, SK. N-Myc knockdown and apigenin treatment controlled growth of malignant neuroblastoma cells having N-Myc amplification. *Gene,* 2013, 529, 27-36.

Hu, XW; Meng, D; Fang, J. Apigenin inhibited migration and invasion of human ovarian cancer A2780 cells through focal adhesion kinase. *Carcinogenesis,* 2008, 29, 2369-2376.

Hui, C; Yujie, F; Lijia, Y; Long, Y; Hongxia, X; Yong, Z; Jungdong, Z; Qianyong, Z; Mantian, M. MicroRNA-34a and microRNA-21 play roles in the chemopreventive effects of 3,6-dihydroxyflavone on 1-methyl-1-nitrosourea-induced breast carcinogenesis. *Breast Cancer Research,* 2012, 14, R80.

Hussain, AR; Khan, AS; Ahmed, SO; Ahmed, M; Platanias, LC; Al-Kuraya, KS; Uddin, S. Apigenin induces apoptosis via downregulation of S-phase kinase-associated protein 2-mediated induction of p27Kip1 in primary effusion lymphoma cells. *Cell Proliferation,* 2010, 43, 170–183.

Ielpo, MT; Basile, A; Miranda, R; Moscatiello, V; Nappo, C; Sorbo, S; Laghi E; Ricciardi, MM; Ricciardi, L; Vuotto, ML. Immunopharmacological properties of flavonoids. *Fitoterapia,* 2000, 71, S101-S109.

Jacobson, KA; Moro, S; Manthey, JA; West, PL; Ji, XD. Interactions of flavones and other phytochemicals with adenosine receptors. *Advances in Experimental Medicine and Biology,* 2002, 505, 163-171.

Jayasooriya, RG; Kang, SH; Kang, CH; Choi, YH; Moon, DO; Hyun, JW; Chang, WY; Kim GY. Apigenin decreases cell viability and telomerases activity in human leukemia cell lines. *Food and Chemical Toxicology,* 2012, 50, 2605-2611.

Jemal, A; Murray, T; Ward, E; Samuels, A; Tiwari, RC; Ghafoor, A; Feuer, EJ; Thun, MJ. Cancer Statistics, CA: *A Cancer Journal for Clinicians,* 2005, 55, 10-30.

Johnson, JL; de Mejia, EG. Flavonoid apigenin modified gene expression associated with inflammation and cancer and induced apoptosis in human pancreatic cancer cells through inhibition of GSK-3β/NF-κB signaling cascade. *Molecular Nutrition and Food Research,* 2013, 57, 2112-2127.

Karmakar, S; Davis, KA; Choudhary, SR; Deeconda, A; Banik, NL; Ray, SK. Bcl-2 inhibitor and apigenin worked synergistically in human malignant neuroblastoma cell lines and increased apoptosis with activation of extrinsic and intrinsic pathways. *Biochemical and Biophysical Research Communications,* 2009, 388, 705-710.

Kelly, EH; Anthony, RT; Dennis, JB. Flavonoid antioxidants: chemistry, metabolism and structure-activity relationships. *Journal of Nutritional Biochemistry,* 2002, 13, 572–584.

Kilani-Jaziri, S; Frachet, V; Bhouri, W; Ghedira, K; Chekir-Ghedira, L; Ronot, X. Flavones inhibit the proliferation of human tumor cancer cell lines by inducing apoptosis. *Drug and Chemical Toxicology*, 2012, 35, 1-10.Kim, HP; Son, KH; Chang, HW; Kang, SS. Anti-inflammatory plant flavonoids and cellular action mechanisms. *Journal of Pharmacological Sciences*, 2004, 96, 229-245.

Kim, SH; Kang, JG; Kim, CS; Ihm, SH; Choi, MG; Yoo, HJ; Lee, SJ. Apigenin induces c-Myc-mediated apoptosis in FRO anaplastic thyroid carcinoma cells. *Molecular and Cellular Endocrinology*, 2013, 369, 130-139.

King, JC; Lu, QY; Li, G; Moro, A; Takahashi, H; Chen, M; Go, VL; Reber, HA; Eibl, G; Hines, OJ. Evidence for activation of mutated p53 by apigenin in human pancreatic cancer. *Biochimica et Biophysica Acta,* 2012, 1823, 593-604.

Kioi, M; Takahashi, S; Kawakami, M; Kawakami, K; Kreitman, RJ; Puri, RK. Expression and targeting of interleukin-4 receptor for primary and advanced ovarian cancer therapy. *Cancer Research,* 65, 8388–8396.

Kuo, CH; Weng, BC; Wu, CC; Yang, SF; Wu, DC, Wang, YC. Apigenin has anti-atropic gastritis and anti-gastric cancer progression effects in Helicobacter pylori-infected Mongolian gerbils. *Journal of Ethnopharmacology,* 2014, 151, 1031-1039.

Lakshmanan, A; Doseff, AI; Ringel, MD; Saji, M; Rousset, B; Zhang, X; Jhiang, SM. Apigenin in combination with Akt inhibition significantly enhances thyrotropin-stimulated radioiodide accumulation in thyroid cells. *Thyroid,* 2014, 24, 878-887.

Lee, Y; Sung, B; Kang, YJ; Kim, DH; Jang, JY; Hwang, SY; Kim, M; Lim, HS; Yoon, JH; Chung, HY; Kim, ND. Apigenin-induced apoptosis is enhanced by inhibition of autophagy formation in HCT116 human colon cancer cells. *International Journal of Oncology,* 2014, 55, 1599-1606.

Lee, SH; Ryu, JK; Lee, KY; Woo, SM; Park, JK; Yoo, JW; Kim, YT; Yoon, YB. Enhanced anti-tumor effect of combination therapy with gemcitabine and apigenin inpancreatic cancer. *Cancer Letters,* 2008, 259, 39-49.

Lefort, ÉC; Blay, J. The dietary flavonoid apigenin enhances the activities of the anti-metastatic protein CD26 on human colon carcinoma cells. *Clinical and Experimental Metastasis,* 2011, 28, 337-349.

Lefort, ÉC; Blay, J. Apigenin and its impact on gastrointestinal cancers. *Molecular Nutrition and Food Research,* 2013, 57, 126-144.

Li, B; Birt, DF. In vivo and in vitro percutaneous absorption of cancer preventive flavonoid apigenin in different vehicles in mouse skin. *Pharmaceutical Research*, 1996, 13, 1710–1715.

Lide, DR. ed. (1997) CRC Handbook of Chemistry and Physics (78th Ed. pp. 3-78), New York, USA, *CRC Press.*

Liu, J; Cao, XC; Xiao, Q; Quan, MF. Apigenin inhibits HeLa sphere-forming cells through inactivation of casein kinase 2α. *Molecular Medicine Reports,* 2014, doi: 10.3892/mmr.2014.2720. (Epub ahead of print)

Long, X; Fan, M; Bigsby, RM; Nephew, KP. Apigenin inhibits antiestrogen-resistant breast cancer cell growth through estrogen receptor-alpha-dependent and estrogen receptor-alpha-independent mechanisms. *Molecular Cancer Therapeutics,* 2008, 7, 2096-2108.

Losi, G; Puia, ; Garzon, G; de Vuona, MC; Baraldi, M. Apigenin modulates GABAergic and glutamatergic transmission in cultured cortical neurons. *European Journal of Pharmaology,* 2004, 205, 41-46.

Lu, HF; Chie, YJ; Yang, MS; Lee, CS; Fu, JJ; Yang, JS; Tan, TW; Wu, SH; M, YS; Ip, SW; Chung, JG. Apigenin induces caspase-dependent apoptosis in human lung cancer A549 cells through Bax-and Bcl-2-triggered mitochondrial pathway. *International Journal of Oncology,* 2010, 36, 1477-1484.

Luo, H; Jiang, BH; King, SM; Chen, YC. Inhibition of cell growth and VEGF expression in ovarian cancer cells by flavonoids. *Nutrition and Cancer,* 2008, 60, 800-809.

Mafuvadze, B; Benakanakere, I; López Pérez, FR; Besch-Williford, C; Ellersieck, MR; Hyder, SM. Apigenin prevents development of medroxyprogesterone acetate-accelerated 7,12-dimethylbenz9a)anthracene induced mammary tumors in Sprague-Dawley rats. *Cancer Prevention Research,* 2011, 4, 1316-1324.

Mafuvadze, B; Cook, M; Xu, Z; Besch-Williford, CL; Hyder, SM. Effects of dietary apigenin on tumor latency, incidence and multiplicity in a medroxyprogesterone acetate-accelerated 7,12-dimenthibenz(a) anthracene-induced breast cancer model. *Nutrition and Cancer,* 2013, 65, 1184-1191.

Mafuvadze, B; Liang, Y; Besch-Williford, CL; Zhang, X; Hyder, SM. Apigenin induces apoptosis and blocks growth of medroxyprogesterone acetate-dependent BT-474 xenograft tumors. *Hormones and Cancer,* 2012, 3, 160-171.

Mahbub, AA; Le Maitre, CL; Haywood-Small, SL; McDougall, GJ; Cross, NA; Jordan-Mahy, N. Differential effects of polyphenols on proliferation and apoptosis in human myeloid and lymphoid leukemia cell lines. *Anti-cancer Agents in Medicinal Chemistry,* 2013, 13, 1601-1613.

Mahomoodally, MF; Gurib-Fakim, A; Subratty AH. Antimicrobial activities and phytochemical profiles of endemic medicinal plants of Mauritius. *Pharmaceutical Biology,* 2005, 43, 237–242.

Manach, C; Scalbert, A; Morand, C; Remesy, C; Jimenez, L. Polyphenols: food sources and bioavailability. *The American Journal of Clinical Nutrition,* 2004, 79, 727-747.

McCann, SE; Freudenheim, JL; Marshall, JR; Graham, S. Risk of human ovarian cancer is related to dietary intake of selected nutrients, phytochemicals and food groups. *The Journal of Nutrition,* 2003, 133, 1937–1942.

McKay, DL; Blumberg, JB. A review of the bioactivity and potential health benefits of chamomile tea (*Matricaria recutita* L.). *Phytotherapy Research,* 2006, 20, 519-530.

McVean, M; Xiao, H; Isobe, K; Pelling, JC. Increase in wild-type p53 stability and transactivational activity by the chemopreventive agent apigenin in keratinocytes. *Carcinogenesis,* 2000, 21, 633–639.

Melstrom, LG; Salabat, MR; Ding, XZ; Milam, BM; Strouch, M; Pelling, JC; Bentrem, DJ. Apigenin inhibits the GLUT-1 glucose transporter and the phosphoinositide 3-kinase/Akt pathway in human pancreatic cancer cells. *Pancreas,* 2008, 37, 426-431.

Melstrom, LG; Salabat, MR; Ding, XZ; Milam, BM; Strouch, M; Pelling, JC; Bentrem, DJ. Apigenin inhibits the GLUT-1 glucose transporter and the phosphoinositide 3-kinase/Akt pathway in human pancreatic cancer cells. *Pancreas*, 2008, 37, 426–431.

Melstrom, LG; Salabat, MR; Ding, XZ; Strouch, MJ; Grippo, PJ; Mirzoeva, S; Pelling, JC; Bentrem, DJ. Apigenin down-regulates the hypoxia response genes: HIF-1α, GLUT-1, and VEGF in human pancreatic cancer cells. *The Journal of Surgical Research,* 2011, 167, 173-181.

Miean, KH; Mohamed, S. Flavonoid (myricetin, quercetin, kaempferol, luteolin, and apigenin) content of edible tropical plants. *Journal of Agricultural and Food Chemistry,* 2001, 49, 3106-3112.

Mirzoeva, S; Kim, ND; Chiu, K; Franzen, CA; Bergan, RC; Pelling, JC. Inhibition of HIF-1 alpha and VEGF expression by the chemopreventive bioflavonoid apigenin is accompanied by Akt inhibition in human prostate carcinoma PC3-N cells. *Molecular carcinogenesis,* 2008, 47, 686-700.

Mohan, N; Ai, W; Chakrabarti, M; Banik, NL; Ray, SK. KLF4 overexpression and apigenin treatment down regulated anti-apoptotic Bcl-2 proteins and matrix metalloproteinases to control growth of human malignant neuroblastoma SK-N-DZ and IMR-32 cells. *Molecular Oncology,* 2013, 7, 464-474.

Mohan, N; Banik, NL; Ray, SK. Combination of N-(4-hydroxyphenyl) retinamide and apigenin suppressed starvation-induced autophagy and promoted apoptosis in malignant neuroblastoma cells. *Neuroscience Letters*, 2011, 502, 24-29.

Murali, R; Soslow, RA; Weigelt, B. Classification of endometrial carcinoma: more than two types. *The Lancet Oncology,* 2014, 15, 268-278.

Nagaprashantha, LD; Vatsyayan, R; Singhal, J; Fast, S; Roby, R; Awasthi, S; Singhal, SS. Anti-cancer effects of novel flavonoid vicenin-2 as a single agent and in synergistic combination with docetaxel in prostate cancer. *Biochemical Pharmacology,* 2011, 82, 1100-1109.

Ni, F; Gong, Y; Li, L; Abdolmaleky, HM; Zhou, JR. Flavonoid ampelopsin inhibits the growth and metastasis of prostate cancer in vitro and in mice. *PLoS One,* 2012, 7, e38802.

O'Toole, SA; Sheppard, BL; Sheils, O; O'Leary, JJ; Spengler, B; Christoffel, V. Analysis of DNA in endometrial cancer cells treated with phytoestrogenic compounds using comparative genomic hybridisation microarrays. *Planta Medica,* 2005, 71, 435–439.

O'Neil, MJ. ed. (2006) The Merck Index: An Encyclopedia of Chemicals, Drugs, and Biologicals: (14th Ed., pp 118) Whitehouse Station, NJ: *Merck and Co., Inc.*

Pan, MH; Lai, CS; Wu, JC; Ho, CT. Molecular mechanisms for chemoprevention of colorectal cancer by natural dietary compounds. *Molecular Nutrition and Food Research,* 2011, 55, 32-45.

Pan, X; Tan, N; Zeng, G; Zhang, Y; Jia, R. Amentoflavone and its derivatives as novel natural inhibitors of human cathepsin B. *Bioorganic and Medicinal Chemistry,* 2005, 13, 5819-5825.

Pandey, M; Kaur, P; Shukla, S; Abbas, A; Fu, P; Gupta, S. Plant flavone apigenin inhibits HDAC and remodels chromatin to induce growth arrest and apoptosis in human prostate cancer cells: in vitro and in vivo study. *Molecular Carcinogenesis,* 2012, 51, 952-962.

Paredes-Gonzalez, X; Fuentes, F; Su, ZY; Kong, AN. Apigenin reactivates Mrf2 anti-oxidative stress signaling in mouse skin epidermal JB6 P+ cells through epigenetics modifications. *The AAPS Journal*, 2014, 16, 727-735.

Park, JA; Ha, SK; Kang, TH; Oh, MS; Cho, MH; Lee, SY; Park, JH, Kim, SY. Protective effect of apigenin on ovariectomy-induced bone loss in rats. *Life Sciences,* 2008, 82, 1217-1223.

Patel, D; Shukla, S; Gupta, S. Apigenin and cancer chemoprevention: progress, potential and promise. *International Journal of Oncology,* 2007, 30, 233-245.

Plaumann, B; Fritsche, M; Rimpler, H; Brandner, G; Hess, RD. Flavonoids activate wild-type p53. *Oncogene*, 1996, 13, 1605-1614.

Reiss, R; Johnston, J; Tucker, K; Desesso, JM; Keen, CL. Estimation of cancer risks and benefits associated with a potential increased consumption of fruits and vegetables. *Food and Chemical Toxicology,* 2012, 50, 4421–4427.

Ren, HY; Tang, XW. Antiproliferaiton and chemo-sensitization effects of apigenin on human lung cancer cells. *Journal of Zhejiang University Medical Sciences,* 2011, 40, 508-514.

Ross, JA; Kasum, CM. Dietary flavonoids: bioavailability, metabolic effects, and safety. *Annual Review of Nutrition,* 2002, 22, 19-34.

Ruela-de-Sousa, RR; Fuhler, GM; Blom, N; Ferreira, CV; Aoyama, H; Peppelenbosch, MP. Cytotoxicity of apigenin on leukemia cell lines: implications for prevention and therapy. *Cell Death and Disease*, 2010, 1, e19; doi: 10.1038/cddis.2009.18.

Sak, K. Cytotoxicity of dietary flavonoids on different human cancer types. *Pharmacognosy Review*, 2014, 8, 122-146.

Salabat, MR; Melstrom, LG; Strouch, MJ; Ding, XZ; Milam, BM; Ujiki, MB; Chen, C; Pelling, JC; Rao, S; Grippo, PJ; McGarry, TJ; Bentrem, DJ. Geminin is overexpressed in human pancreatic cancer and downregulated by the bioflavonoid apigenin in pancreatic cancer cell lines. *Molecular Carcinogenesis,* 2008, 47, 835-844.

Schröder-van der Elst, JP; van der Heide, D; Romjin, JA; Smit, JW. Differential effects of natural flavonoids on growth and iodide content in a human Na(+)/I(-) symporter-transfected follicular thyroid carcinoma cell line. *European Journal of Endocrinology,* 2004, 150, 557-564.

Seelinger, G; Merfort, I; Wolfle, U; Schempp, CM. Anti-carcinogenic effects of the flavonoid luteolin. *Molecules,* 2008, 13, 2628-2651.

Seo, HS; Choi, HS; Kim, SR; Choi, YK; Woo, SM; Shin, I; Woo, JK; Park, SY; Shin, YC; Ko, SG. Apigenin induces apoptosis via extrinsic pathway, inducing p53 and inhibiting STAT3 and NFκB signaling in HER2-overexpressing breast cancer cells. *Molecular and Cellular Biochemistry,* 2012, 366, 319-334.

Shao, H; Jing, K; Mahmoud, E; Huang, H; Fang, X; Yu, C. Apigenin sensitizes colon cancer cells to antitumor activity of ABT-263. *Molecular Cancer Therapeutics,* 2013, 12, 2640-2650.

Shukla, S; Bhaskaran, N; Babcook, MA; Fu, P; Maclennan, GT; Gupta, S. Apigenin inhibits prostate cancer progression in TRAMP mice via targeting PI3K/Akt/FoxO pathway. *Carcinogenesis,* 2014, 35, 452-460.

Shukla, S; Fu, P; Gupta, S. Apigenin induces apoptosis by targeting inhibitor of apoptosis proteins and Ku70-Bax interaction in prostate cancer. *Apoptosis,* 2014, 19, 883-894.

Shukla, S; Gupta, S. Apigenin suppresses insulin-like growth factor I receptor signaling in human prostate cancer: an in vitro and in vivo study. *Molecular Carcinogenesis,* 2009, 48, 243-252.

Shukla, S; Gupta, S. Apigenin: a promising molecule for cancer prevention. *Pharmaceutical Research,* 2010, 27, 962-978.

Shukla, S; Gupta, S. Apigenin-induced cell cycle arrest is mediated by modulation of MAPK, PI3K-Akt, and loss of cyclin D1 associated retinoblastoma dephosphorylation in human prostate cancer cells. *Cell Cycle,* 2007, 6, 1102-1114.

Shukla, S; MacLennan, GT; Flask, CA; Fu, P; Mishra, A; Resnick, MI; Gupta, S. Blockade of beta-catenin signaling by plant flavonoid apigenin suppresses prostate carcinogenesis in TRAMP mice. *Cancer Research,* 2007, 67, 6925-6935.

Silverberg, SG; Kurman, RJ; Nogales, F; Mutter, GL; Kubik-Huch, RA; Tavassoli, FA. "Epithelial tumours and related lesions," in Pathology & Genetics of Tumours of the Breast and Female Genital Organs, A. Fattaneh, Tavassoli, and P. Devilee, Eds., World Health Organization Classification of Tumours, pp. 221–228, *IARC Press,* Lyon, France, 2003.

Singh, O; Khanam, Z; Misra, N; Srivastava, MK. Chamomile (Matricaria chamomilla L.): An overview. *Pharmacognosy Review*, 2011, 5, 82-95.

Singh, P; Mishra, SK; Neol, S; Sharma, S; Rath, SK. Acute exposure of apigenin induces hepatotoxicity in Swiss mice. *PLoS ONE*, 2012, 7: e31964. doi:10.1371/journal.pone.0031964.

Sporn, MB. Approaches to prevention of epithelial cancer during the preneoplastic period. *Cancer Research,* 1976, 36, 2699 2702.

Srivastava, JK; Gupta, S. Antiproliferative and apoptotic effects of chamomile extract in various human cancer cells. *Journal of Agricultural and Food Chemistry,* 2007, 55, 9470–9478.

Strouch, MJ; Milam, BM; Melstrom, LG; McGill, JJ; Salabat, MR; Ujiki, MB; Ding, XZ; Bentrem DJ. The flavonoid apigenin potentiates the growth inhibitory effects of gemcitabine and abrogates gemcitabine resistance in human pancreatic cancer cells. *Pancreas,* 2009, 38, 409-415.

Surh, YJ. Cancer chemoprevention with dietary phytochemicals. *Nature Reviews Cancer,* 2003, 3, 768-780.

Tang, NP; Zhou, B; Wang, B; Yu, RB; Ma, J. Flavonoids intake and risk of lung cancer: a meta-analysis. *Japanese Journal of Clinical Oncology*, 2009, 39, 352–359.

Todaro, M; Lombardo, Y; Francipane, MG; Alea, MP; Cammareri, P; Iovino, F; Di Stefano, AB; Di Bernardo, C; Agrusa, A; Condorelli, G; Walczak, H; Stassi, G. Apoptosis resistance in epithelial tumors is mediated by tumor-cellderived interleukin-4. *Cell Death and Differentiation*, 2008, 15, 762–772.

Toker, G; Küpeli, E; Memisoglu, M; Yesilada, E. Flavonoids with antinociceptive and anti-inflammatory activities from the leaves of *Tilia argentea* (silver linden). *Journal of Ethnopharmacology,* 2004, 95, 393-397.

Tong, X; Smith, KA; Pelling, JC. Apigenin, a chemopreventive bioflavonoid, induces AMP-activated protein kinase activation in human keratinocytes. *Molecular Carcinogenesis,* 2012, 51, 268-279.

Torkin, R; Lavoie, JF; Kaplan, DR; Yeger, H. Induction of caspase-dependent, p53-mediated apoptosis by apigenin in human neuroblastoma. *Molecular Cancer Therapeutics,* 2005, 4, 1-11.

Turktekin, M; Konac, E; Onen, HI; Alp, E; Yilmaz, A; Menevse, S. Evaluation of the effects of the flavonoid apigenin on apoptotic pathway gene expression on the colon cancer cell line (HT-29). *Journal of Medicinal Food,* 2011, 14, 1107-1117.

Ujiki, MB; Ding, XZ; Salabat, MR; Bentrem, DJ; Golkar, L; Talamonti, MS; Bell, RH Jr; Iwamura, T; Adrian, TE. Apigenin inhibits pancreatic cancer cell proliferation through G2/M cell cycle arrest. *Molecular Cancer,* 2006, 5, 76-83.

Vargo, MA; Voss, OH; Poustka, F; Cardounel, AJ; Grotewold, E; Doseff, AI. Apigenin-induced-apoptosis is mediated by the activation of PKCδ and caspases in leukemia cells. *Biochemical Pharmacology,* 2006, 72, 681-692.

Viola, H; Wasowski, C; Levi de Stein, M; Wolfman, C; Silveira R; Dajas, F; Medina, JH; Paladini, AC. Apigenin, a component of *Matricaria recutita* flowers, is a central benzodiazepine receptors-ligand with anxiolytic effects. *Planta Medica,* 1995, 61, 213-216.

Wang, QR; Yao, XQ; Wen, G; Fan, Q; Li, YJ; Fu, XQ; Li, CK; Sun, XG. Apigenin suppresses the growth of colorectal cancer xenografts via phosphorylation and up-regulated FADD expression. *Oncology Letters,* 2011, 2, 43-47.

Wang, W; Heideman, L; Chung, CS; Pelling, JC; Koehler, KJ; Birt, DF. Cell-cycle arrest at G2/M and growth inhibition by apigenin in human colon carcinoma cell lines. *Molecular Carcinogenesis,* 2000, 28, 102–110.

Wang, W; VanAlstyne, PC; Irons, KA; Chen, S; Stewart, JW; Birt, DF. Individual and interactive effects of apigenin analogs on G2/M cell-cycle arrest in human colon carcinoma cell lines. *Nutrition and Cancer*, 2004, 48, 106–114.

Wang, YC; Huang, KM. In vitro anti-inflammatory effect of apigenin in the *Helicobacter pylori*-infected gastric adenocarcinoma cells. *Food and Chemical Toxicology,* 2013, 53, 376-383.

Wei, H; Tye, L; Bresnick, E; Birt, DF. Inhibitory effect of apigenin, a plant flavonoid, on epidermal ornithine decarboxylase and skin tumor promotion in mice. *Cancer Research,* 1990, 50, 499-502.

Williams, CA; Harborne, JB; Tomas-Barberan, FA. Biflavonoids in the primitive monocots *Isophysis tasmanica* and *Xerophyta plicata. Phytochemistry,* 1987, 26, 2553-2555.

World Cancer Research Fund / American Institute for Cancer Research. Continuous Update Project Report. Food, Nutrition, Physical Activity, and the Prevention of Ovarian Cancer 2014. Available at http://www.dietandcancerreport.org/cup/cup_resources.php

Wu, C; Chen, F; Rushing, JW; Wang, X; Kim, HJ; Huang, G; Haley-Zitlin, V; He, G. Antiproliferative activities of parthenolide and golden feverfew extract against three human cancer cell lines. *Journal of Medicinal Food*, 2006, 9, 55-61.

Wu, DG; Yu, P; Li, JW; Jiang, P; Sun, J; Wang, HZ, Zhang, LD; Wen, MB; Bie, P. Apigenin potentiates the growth inhibitory effects by IKK-β-mediated NK- κB activation in pancreatic cancer cells. *Toxicology Letters,* 2014, 224, 157-164.

Wu, K; Yuan, LH; Xia, W. Inhibitory effects of apigenin on the growth of gastric carcinoma SGC-7901 cells. *World Journal of Gastroenterology,* 2005, 11, 4461-4464.

Yang, CS; Landau, JM; Huang, MT; Newmark, HL. Inhibition of carcinogenesis by dietary polyphenolic compounds. *Annual Review of Nutrition,* 2001, 21, 381-406.

Yin, F; Giuliano, AE; Van Herle, AJ. Growth inhibitory effects of flavonoids in human thyroid cancer cell lines. *Thyroid,* 1999, 9, 369-376.

Yin, F; Giuliano, AE; Van Herle, AJ. Signal pathways involved in apigenin inhibition of growth and induction of apoptosis of human anaplastic thyroid cancer cells. *Anticancer Research,* 1999, 19, 4297-4303.

Zhang, CX; Ho, SC; Chen, YM; Fu, JH; Cheng, SZ; Lin, FY. Greater vegetable and fruit intake is associated with a lower risk of breast cancer among Chinese women. *International Journal of Cancer*, 2009, 125, 181–188.

Zhang, M; Xie, X; Lee, AH; Binns, CW. Soy and isoflavone intake are associated with reduced risk of ovarian cancer in southeast China. *Nutrition and Cancer,* 2004, 49, 125–130.

Zhao, R; Xiang, N; Domann, FE; Zhong, W. Effects of selenite and genistein on G2/M cell cycle arrest and apoptosis in human prostate cancer cells. *Nutrition and Cancer,* 2009, 61, 397-407.

Zheng, PW; Chiang, LC; Lin, CC. Apigenin induced apoptosis through p53-dependent pathway in human cervical carcinoma cells. *Life Sciences,* 2005, 76, 1367–1379.

Zhong, Y; Krisanapun, C; Lee, SH; Nualsanit, T; Sams, C; Peungvicha, P; Baek, SJ. Molecular targets of apigenin in colorectal cancer cells: involvement of p21, NAG-1 and p53. *European Journal of Cancer,* 2010, 46, 3365-3374.

Zhu, Y; Wu, J; Li, S; Wang, X; Liang, Z; Xu, X; Xu, X; Hu, Z; Lin, Y; Chen, H; Qin, J; Mao, Q; Xie, L. Apigenin inhibits migration and invasion via modulation of epithelial mesenchymal transition in prostate cancer. *Molecular Medicine Reports,* 2014; doi: 10.3892/mmr.2014.2801

In: Apigenin and Naringenin
Editor: Nilus M. Stacks

ISBN: 978-1-63463-987-3
© 2015 Nova Science Publishers, Inc.

Chapter 8

NARINGENIN REGULATES THE INTESTINAL TIGHT JUNCTION BARRIER AND INFLAMMATION

Takuya Suzuki*
Department of Biofunctional Science and Technology,
Graduate School of Biosphere Science, Hiroshima University, Japan

ABSTRACT

A primary function of gastrointestinal epithelial cells is to provide a physical barrier to the diffusion of pathogens, toxins, and allergens from the external environment into the tissues. Defects in the barrier function resulting in the permeation of luminal inflammatory substances induce an abnormally robust inflammatory response. It has been reported that barrier defects play a crucial role in the pathogenesis of various intestinal diseases, such as inflammatory bowel disease, irritable bowel syndrome, and food allergies. Therefore, the maintenance of the intestinal barrier is imperative for intestinal mucosal homeostasis and our health. The intestinal barrier is maintained through the interactions of several barrier components, including the adhesive mucous gel layer, the mucosal immune system, and intercellular tight junctions (TJs). The TJ is a multiple protein complex, which locates around the apical end of the lateral membrane of polarized epithelial cells and selectively regulates the paracellular passage of ions, molecules and water. Although the TJ structure has been elucidated on a molecular level in recent studies, less information is available on the roles of food factors on intestinal barrier regulation. Naringenin is a major polyphenol isolated from citrus fruits. Recently, we have shown that naringenin enhances and protects TJ barrier integrity in intestinal cells and a murine model of colitis. Accordingly, naringenin appears to participate in the physiological and pathological regulation of the intestinal TJs. This chapter presents the recent advances in our understanding of the naringenin-mediated regulation of the intestinal TJ barrier.

* Address correspondence to: Takuya Suzuki, Ph. D., 1-4-4, Kagamiyama, Higashi-Hiroshima, 739-8528, Japan. TEL: 81-82-424-7984. Fax: 81-82-424-7916; E-mail: takuya@hiroshima-u.ac.jp.

INTESTINAL BARRIER AND DISEASES

Intestinal mucosal surfaces are lined with epithelial cells, which are responsible for nutrient and water absorption while establishing a physical barrier between the hostile external environment and the internal milieu. In healthy individuals, the intestinal barrier effectively restricts the diffusion of luminal proinflammatory molecules, such as microorganisms, pathogens, toxins, and allergens, into the mucosal tissues and circulatory system (Figure 1); however, impairment of the barrier resulting in the invasion of these molecules abnormally activates the mucosal immune system and induces chronic inflammation. Therefore, intestinal barrier defects are closely involved in the pathogenesis of several intestinal and systemic inflammatory disorders, such as inflammatory bowel diseases (IBDs), celiac disease, food allergy, metabolic syndrome, and alcoholic liver diseases. Among these disorders, IBD pathogenesis has been well-studied in relation to intestinal barrier defects. IBDs, including Crohn's disease (CD) and ulcerative colitis (UC), are chronic and relapsing-remitting inflammatory diseases. The patients suffer from a variety of symptoms including abdominal cramps, bloody diarrhea, and fever. Although IBD pathogenesis remains unclear and is possibly associated with multiple factors, evidence from basic and clinical studies suggests that intestinal barrier defects can initiate IBDs, with intestinal hyperpermeability observed in both the non-involved intestinal segments as well as the involved segments [2]. A subset of first-degree relatives of patients with CD have shown increased intestinal permeability despite being apparently healthy [3]. Further investigations are definitely required to understand the role of intestinal barrier defects in the pathogenesis of a variety of diseases, including IBDs; however, the promotion and protection of intestinal barrier integrity could afford a novel approach to the prevention and amelioration of these diseases.

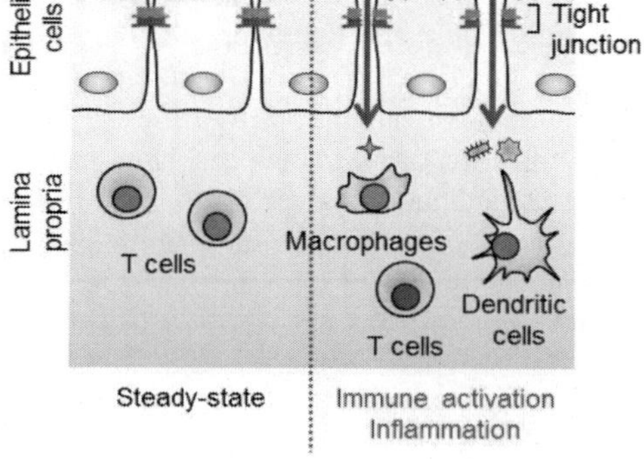

Figure 1. Barrier function of intestinal tight junctions. The intestinal epithelium provides a physical barrier to luminal inflammatory substances. Impaired barrier function allows the diffusion of these inflammatory substances into mucosal tissues and the circulatory system, resulting in chronic activation of the immune system in tissues.

MOLECULAR BASIS OF INTESTINAL TIGHT JUNCTIONS

The intestinal barrier is maintained through the interactions of different components, including the mucus layer, immunoglobulin A, antimicrobial peptides, and intercellular tight junctions (TJs). The TJ is a multiple protein complex consisting of transmembrane and cytosolic proteins. It locates on the apical portion of lateral membranes of epithelial cells and regulates the paracellular passage of solutes and water (Figure 2). The TJ structure provides a selectively permeable barrier that limits the permeation of luminal hostile molecules, such as microorganisms and dietary antigens, while allowing the appropriate absorption of nutrients and water via paracellular pathways. To date, 4 types of transmembrane proteins, claudins, occludin, junctional adhesion molecules (JAMs), and tricellulin, have been identified [4-7]. The intracellular domains of these transmembrane proteins are directly or indirectly associated with cytosolic plaque proteins, including zonula occludens (ZO) proteins, which in turn anchor the TJ complexes to perijunctional actin cytoskeletal rings within the cells. Therefore, the cytoskeletal association of the TJ complexes is essential to the maintenance of TJ barrier integrity [8].

Claudins, occludin, and tricellulin are tetratranspanin membrane proteins in TJ structures, and show 2 extracellular loops in the paracellular space. Interactions between the extracelluar loops from adjacent cells form a selective barrier in paracellular pathways. Claudins form a multigene family with at least 24 members in human and mice. Each isoform shows a unique expression pattern in tissues and cell lines. In mouse intestines, claudin-1, -2, -3, -4, -5, -7, -8, -10, -12, -13,-14, -15, -17, and -18 are detected at the gene expression level and the expression level of each isoform varies according to the intestinal segment [9]. Despite their structural similarity, claudins exhibit different properties and are roughly classified into 2 types; barrier-forming and pore-forming isoforms. In the intestines, claudin-1, -3, -4, -5, -8, -9, -11, and -14 are categorized as barrier-forming claudins, while claudin-2, -7, -12, and -15 are pore-forming claudins. Occludin was the first integral membrane protein identified in TJ structures. The homophilic interaction between the occludin in adjacent cells appears to form a barrier against macromolecules, but not against small ions [10]. Tricellulin preferentially localizes at the tricellular junctions, where 3 adjacent cells create cell-cell contacts, although it is also observed at bicellular junctions created by 2 adjacent cells. The amino acid sequence of C-terminal is ~32% identical to that of occludin, while the N-terminal domain is longer than that of occludin. Recent studies have demonstrated that tricellulin plays an important role in the epithelial TJ barrier, both at tricellular and bicellular junctions [11]. The JAM family belongs to the immunoglobulin superfamily and is characterized by the existence of a transmembrane domain. In intestinal epithelial cells, JAM-A, JAM-4, and CAR (Coxsackievirus and adenovirus receptor) are expressed and reportedly contribute to the TJ barrier [12-14]. Cytosolic plaque proteins, such as ZO proteins, seem to provide a scaffold for the assembly of these transmembrane proteins in the TJ structure. Further, a recent study has elegantly shown that ZO proteins also have an important role in the assembly of occludin and claudins at cell-cell junctions [15, 16].

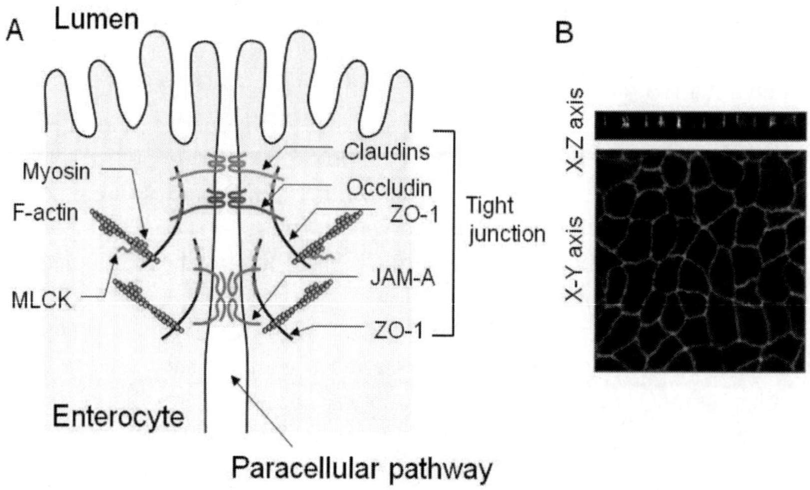

Figure 2. Molecular structure of intestinal tight junctions. (A) The TJ is a multiple protein complex consisting of transmembrane and cytosolic proteins that acts to prevent the paracellular diffusion of noxious substances into tissues. The TJ complex is directly associated with the perijunctional actin cytoskeletal rings. (B) Confocal images of an immunolabeled TJ protein, zonula occludens (ZO)-1, collected from human intestinal Caco-2 cells. An immunofluorescence image of ZO-1 indicating the localization of the TJ protein complex at the apical portion of the lateral membrane of epithelial cells.

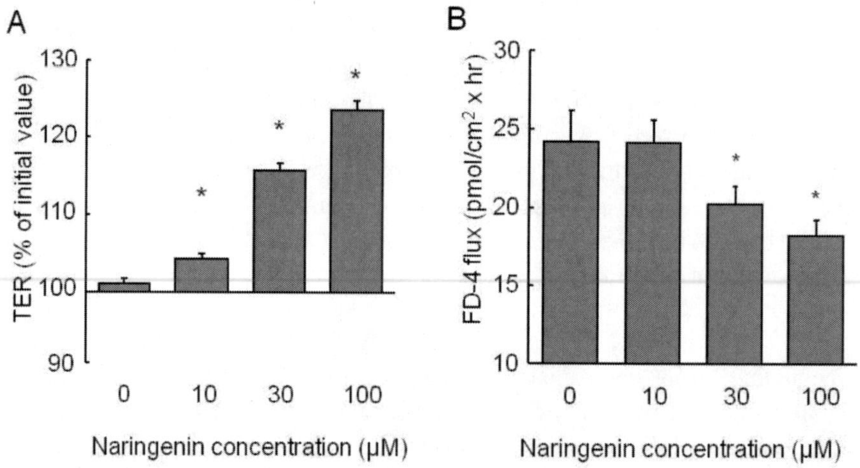

Figure 3. Effects of naringenin on TJ barrier integrity in intestinal cells. Human intestinal Caco-2 cells were incubated with naringenin (0~100 μM) for 48 hr and 2 indicators of TJ barrier integrity, transepithelial electrical resistance (TER, A) and FITC-dextran flux (B), were measured. Naringenin increased the TER and decreased FITC-dextran flux in a dose-dependent manner. Values are means ± SEM, n = 4. *Different from the control value, $p < 0.05$.

INTESTINAL TJ BARRIER REGULATION BY NARINGENIN

Recently, dietary polyphenols have been attracting a good deal of attention due to their health promoting effects. As intestinal barrier function is closely related to our health, as described above, the effects of some polyphenols on TJ integrity have been evaluated in human intestinal Caco-2 cells [17]. Naringenin, a citrus polyphenol, promotes intestinal TJ barrier integrity, as demonstrated by the increased transepithelial electrical resistance (TER), an indicator of TJ barrier integrity. Another study has shown that naringenin suppresses dextran (4.4 kDa) flux through the TJ pathways in a dose-dependent manner [18] (Figure 3). These studies were the first to report the enhancing effect of naringenin on intestinal TJ barrier integrity.

The TJ structure consists of transmembrane and cytosolic proteins, such as claudin, occludin and ZOs, and their expression and cytoskeletal association basically determine TJ barrier integrity. Naringenin increases the total expression of claudin-4 as well as the cytoskeletal association of certain TJ proteins, such as claudin-1, occludin, and ZO-2 (Figure 4). Increased junctional localization of occludin, claudin-1 and claudin-4 have also been observed in immunofluorescence microscopic images (Figure 5). With regard to claudin-4,

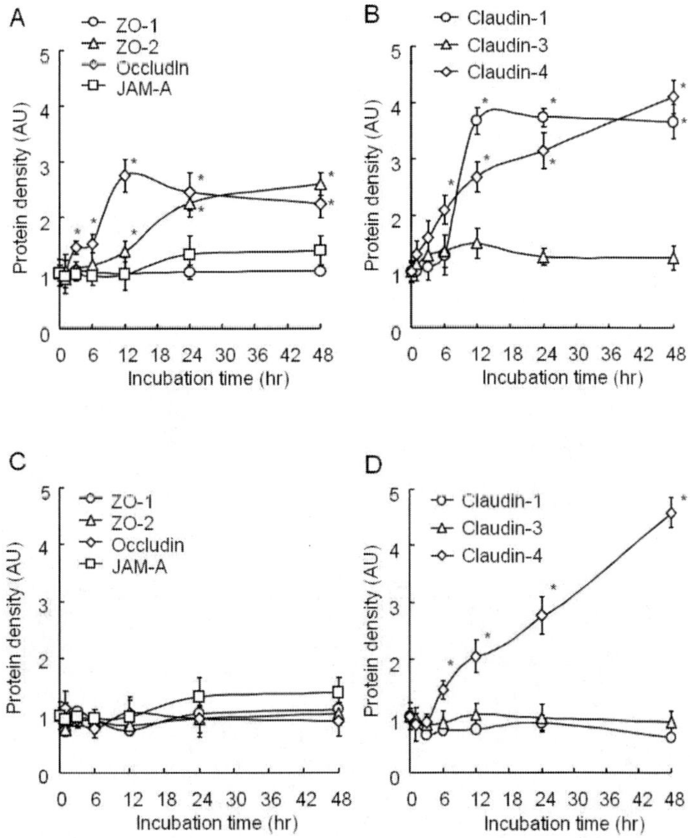

Figure 4. Effects of naringenin on the total expression and cytoskeletal association of TJ proteins in intestinal cells. Human intestinal Caco-2 cells were incubated with naringenin (100 μM). Cytoskeletal fractions (A and B) and whole cell extracts (C and D) were prepared at different time points. TJ proteins, ZO-1, ZO-2, occludin, JAM-A, claudin-1, claudin-3, and claudin-4, in each fraction were immunoblotted and the specific bands for each protein were quantified by densitometric analysis. The density values were normalized against the values before incubation. Naringenin increased the cytoskeletal association of ZO-2, occludin, claudin-1, and claudin-4 as well as the total expression of claudin-4. Values are means ± SEM, n = 4. *Different from the value before incubation, $p < 0.05$.

naringenin is known to enhance the mRNA expression and promoter activity of claudin-4, indicating that naringenin induces the claudin-4 expression at a transcriptional level (Figure 6). Gene transcription is often controlled by the binding of transcription factors to specific regions of the gene promoter sequences, and a previous study has demonstrated that a transcription factor, sp1, is a key regulator of claudin-4 transcription [19]. Site-directed mutation of sp1 in the claudin-4 promoter sequence partially attenuated the naringenin-mediated activation of the claudin-4 promoter, indicating the involvement of sp1 in naringenin-induced claudin-4 expression (Figure 7). In addition, the pharmacological inhibition of sp1 also suppressed naringenin-mediated claudin-4 promoter activation.

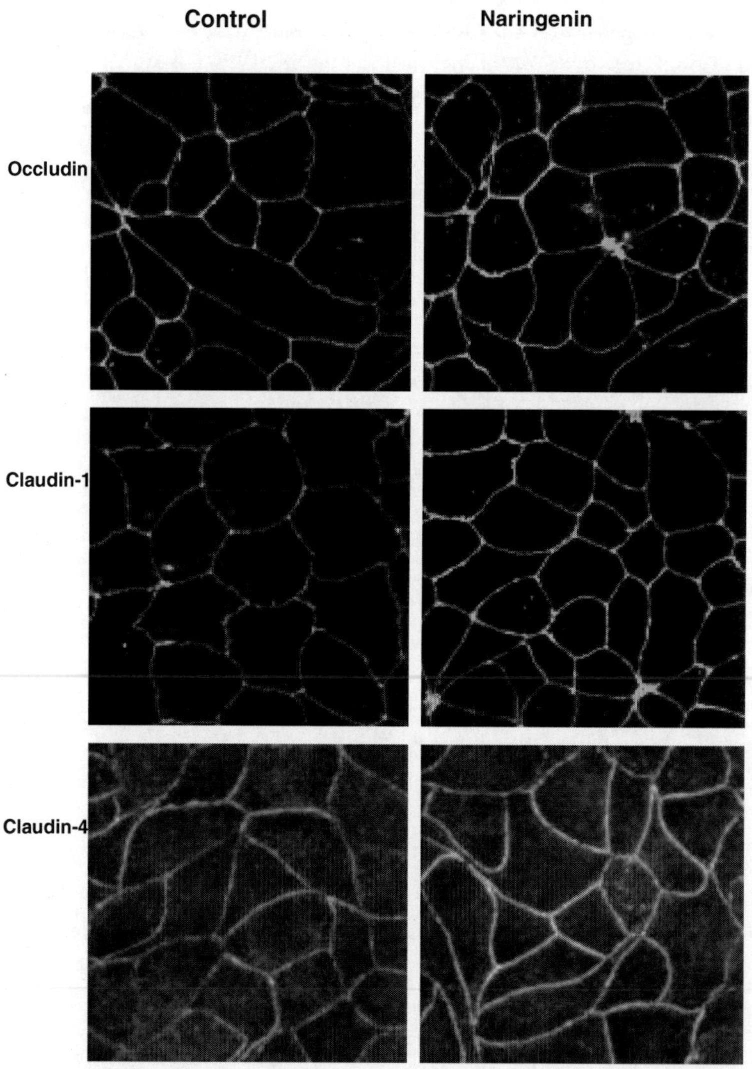

Figure 5. Effects of naringenin on the localization of TJ proteins in intestinal cells. Human intestinal Caco-2 cells were incubated without or with naringenin (100 μM) for 48 hr, and occludin, claudin-1, and claudin-4 were immunostained. Images were collected by confocal microscopy. Naringenin enhanced the immunolocalization of occludin, claudin-1, and claudin-4 at the junctional regions.

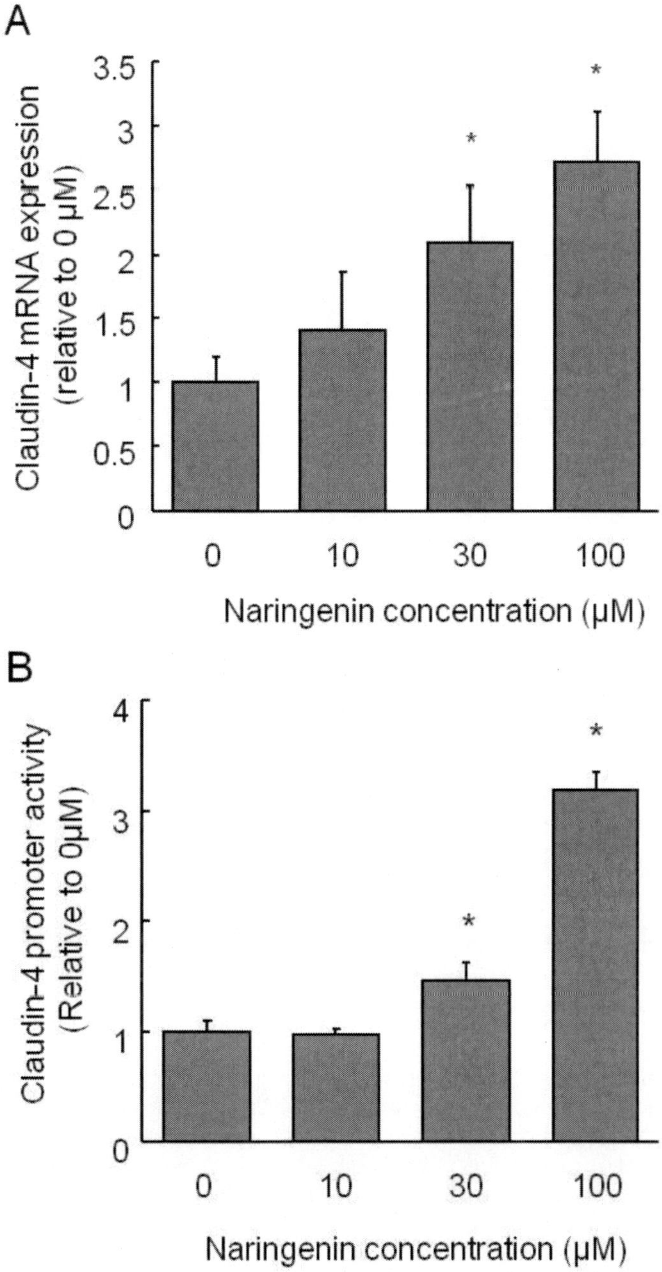

Figure 6. Effects of naringenin on the mRNA expression and promoter activity of claudin-4 in intestinal cells. (A) Human intestinal Caco-2 cells were incubated with naringenin (0~100 μM) for 48 hr and claudin-4 mRNA expression was analyzed by quantitative RT-PCR analysis. Naringenin increased claudin-4 mRNA expression in a dose-dependent manner. (B) Caco-2 cells transfected with a pGL3 vector carrying a claudin-4 promoter sequence were incubated with naringenin (0~100 μM) for 48 hr and luciferase activity, indicating the claudin-4 promoter activity, was measured. Naringenin (30 and 100 μM) increased the claudin-4 promoter activity in cells. Values are means ± SEM, n = 4. *Different from the control value, $p < 0.05$.

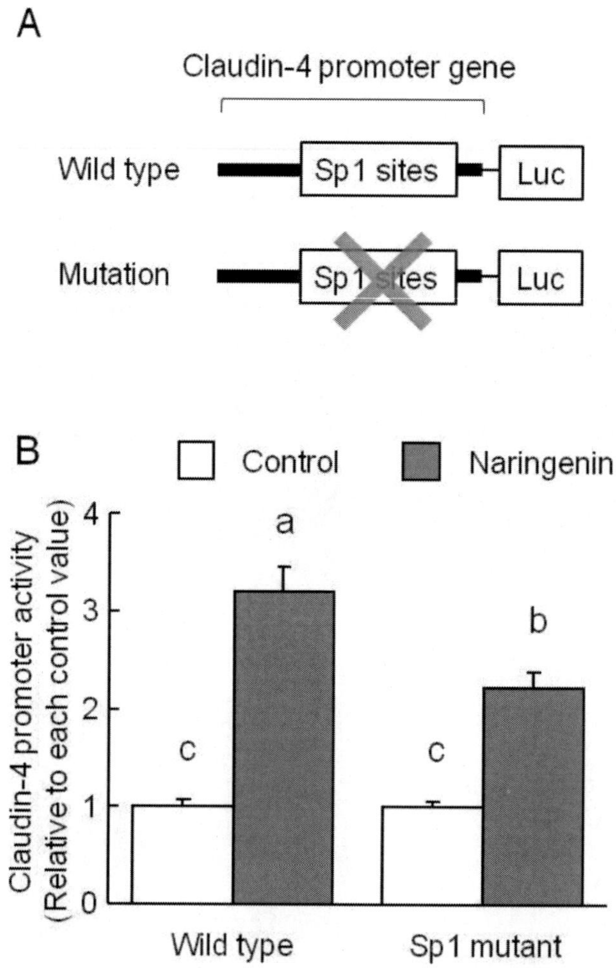

Figure 7. Effect of sp1 binding site mutations on naringenin-mediated claudin-4 promoter activation. (A) A schematic drawing of the claudin-4 promoter gene in a luciferase reporter plasmid is shown. Sp1 binding sites in the claudin-4 promoter gene were mutated in the plasmids. (B) Human intestinal Caco-2 cells transfected with the claudin-4 promoter plasmids (wild type and sp1 mutated) were incubated without or with naringenin (100 μM) for 48 hr and luciferase activity, indicating the claudin-4 promoter activity, was measured. The Sp1 mutation partially suppressed the naringenin-mediated claudin-4 promoter activation. Values are means ± SEM, n = 4. *Means without a common letter significantly differ, $p < 0.05$.

The molecular mechanisms underlying the naringenin-mediated increase in the cytoskeletal association of TJ proteins remain unclear; however, occludin phosphorylation seems to be involved in the regulatory mechanism. Based on the results of previous studies, occludin is thought to be highly phosphorylated on threonine and serine residues in the intact epithelium and its phosphorylation status plays a role in both occludin assembly and TJ integrity [20-22]. Interestingly, immunoblot analysis has demonstrated that naringenin-mediated occludin assembly occurs in a similar manner to occludin phosphorylation. Taken

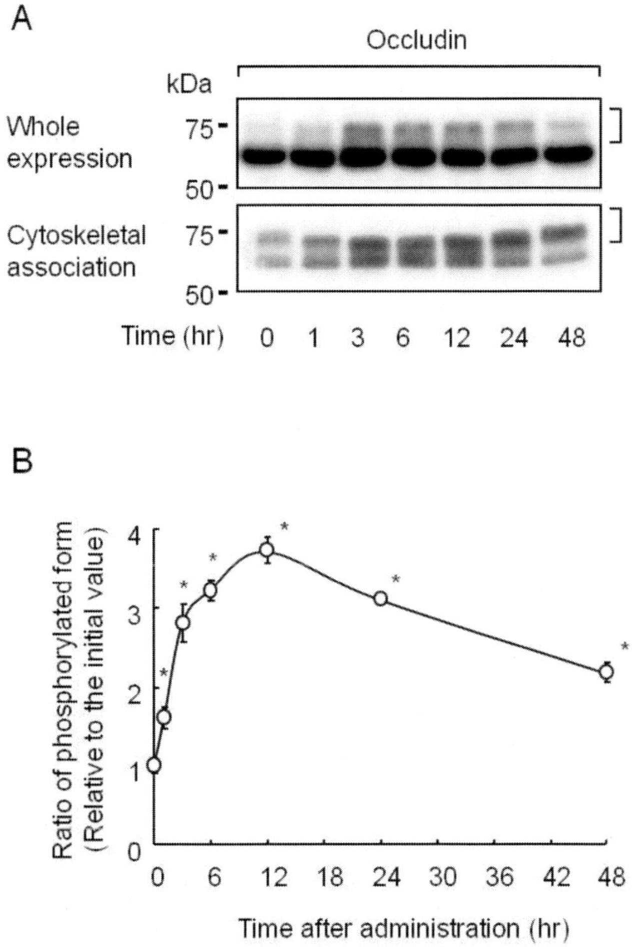

Figure 8. Effects of naringenin on the phosphorylation and cytoskeletal association of occludin in intestinal cells. Human intestinal Caco-2 cells were incubated with naringenin (100 μM). Whole cell extracts and cytoskeletal fractions were prepared at different time points. Occludin in each fraction was immunoblotted (A). #The hyper-phosphorylated form of occludin. The ratio of the hyper-phosphorylated form (upper bands) to the total expression was quantified by densitometric analysis (B). The ratios were normalized against the values before incubation. Naringenin increased occludin phosphorylation in a similar manner to its cytoskeletal association. Values are means ± SEM, n = 4. *Different from the value before incubation, $p < 0.05$.

together, these findings suggest that naringenin induces occludin assembly at TJs via its phosphorylation (Figure 8). Protein phosphorylation levels are balanced by activities of various kinases and phosphatases within cells. Some kinases, including protein kinase C (PKC) η and ζ, and casein kinase (CK) I and II, have been found to phosphorylate occludingand regulate TJ barrier integrity [20, 23-25]. On the other hand, it has been reported that protein phosphatase 1 and 2A dephosphorylate occludin and negatively regulate barrier integrity [26, 27]. These studies suggest that naringenin affects the activities of one or another of these kinases and phosphatases, resulting in increased occludin phosphorylation.

PREVENTIVE ROLE OF NARINGENIN ON INTESTINAL BARRIER DEFECTS AND INFLAMMATION

Intestinal barrier defects are closely related to intestinal disorders, including IBD, as described above. In basic scientific studies, different rodent models of colitis have been developed using dextran sulfate sodium (DSS) and acetic acid, and their symptoms closely resemble those of human IBD. The supplemental feeding of naringenin to mice suppressed clinical colitis scores based on the severity of body weight loss, diarrhea and bloody stool induced by DSS (Figure 9) [28, 29]. A similar effect was also observed in rats with acetic

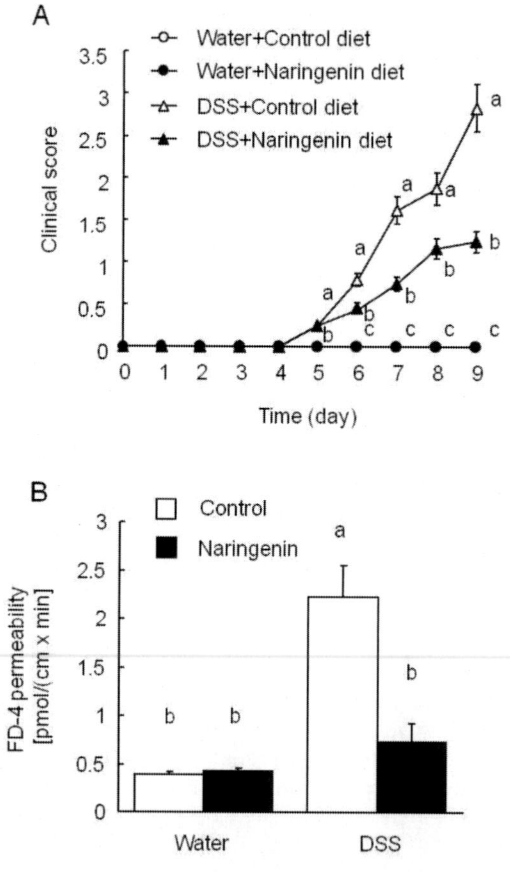

Figure 9. Effect of naringenin on the clinical score and intestinal barrier function in colitic mice. Mice were administered water or 2% dextran sulfate sodium (DSS) in combination with the feeding of control or naringenin-containing diets for 9 days. (A) The clinical colitis score was evaluated based on diarrhea, bloody stool, and body weight loss. The DSS administration gradually increased the clinical score, indicating the development of colitis, on and after day 5 whereas naringenin feeding partially suppressed this increase. Values are means ± SEM, n = 7. Means at a time point without a common letter differ, $p < 0.05$. (B) On day 9, FITC-dextran permeability in the colon of mice was measured using everted colonic segments. DSS administration increased FITC-dextran (FD-4) permeability whereas naringenin feeding effectively suppressed this increase. Values are means ± SEM, n = 7. Means without a common letter significantly differ, $p < 0.05$.

acid-induced colitis [30]. Feeding naringenin also suppressed the colonic expression of proinflammatory cytokines such as TNF-α, IL-6, and IL-17A. This anti-colitic effect of naringenin is speculated to result from the impact of naringenin on both immune and epithelial cells in the colon. A study using mouse macrophages showed that naringenin directly inhibits the TNF-α-induced activation of NF-κB, which is known to mediate a key signaling pathway involved in intestinal inflammation [28]. Feeding naringenin protects the expression and localization of TJ proteins, including occludin and claudin-3, in the colon of colitic mice, leading to the restoration of intestinal TJ barrier function [29]. Although the clinical use of naringenin must be investigated in future studies, it appears that naringenin could be an effective therapeutic and preventive tool for IBD patients.

CONCLUSION

In developing our knowledge of the molecular structure and physiological and pathological roles of epithelial TJs, the effects of nutritional factors on the intestinal TJ barrier have attracted a good deal of attention. In *vivo* and *vitro* studies have demonstrated that naringenin promotes and protects the intestinal TJ barrier, which plays an essential role in our health. Although the molecular mechanisms underlying the naringenin-mediated effects as well as the use of naringenin in humans should be investigated in future studies, the barrier effect of naringenin could be applied to the development of therapeutic and preventive tools against the intestinal TJ-related disorders.

REFERENCES

[1] Farhadi, A., Banan, A., Fields, J., Keshavarzian, A., Intestinal barrier: an interface between health and disease. *J Gastroenterol Hepatol* 2003, *18*, 479-497.

[2] Soderholm, J. D., Peterson, K. H., Olaison, G., Franzen, L. E., et al., Epithelial permeability to proteins in the noninflamed ileum of Crohn's disease? *Gastroenterology* 1999, *117*, 65-72.

[3] Hollander, D., Vadheim, C. M., Brettholz, E., Petersen, G. M., et al., Increased intestinal permeability in patients with Crohn's disease and their relatives. A possible etiologic factor. *Ann Intern Med* 1986, *105*, 883-885.

[4] Furuse, M., Hirase, T., Itoh, M., Nagafuchi, A., et al., Occludin: a novel integral membrane protein localizing at tight junctions. *J Cell Biol* 1993, *123*, 1777-1788.

[5] Furuse, M., Fujita, K., Hiiragi, T., Fujimoto, K., Tsukita, S., Claudin-1 and -2: novel integral membrane proteins localizing at tight junctions with no sequence similarity to occludin. *J Cell Biol* 1998, *141*, 1539-1550.

[6] Ikenouchi, J., Furuse, M., Furuse, K., Sasaki, H., Tsukita, S., Tricellulin constitutes a novel barrier at tricellular contacts of epithelial cells. *J Cell Biol* 2005, *171*, 939-945.

[7] Martin-Padura, I., Lostaglio, S., Schneemann, M., Williams, L., et al., Junctional adhesion molecule, a novel member of the immunoglobulin superfamily that distributes at intercellular junctions and modulates monocyte transmigration. *J Cell Biol* 1998, *142*, 117-127.

[8] Rao, R. K., Basuroy, S., Rao, V. U., Karnaky Jr, K. J., Gupta, A., Tyrosine phosphorylation and dissociation of occludin-ZO-1 and E-cadherin-beta-catenin complexes from the cytoskeleton by oxidative stress. *Biochem J* 2002, *368*, 471-481.

[9] Holmes, J. L., Van Itallie, C. M., Rasmussen, J. E., Anderson, J. M., Claudin profiling in the mouse during postnatal intestinal development and along the gastrointestinal tract reveals complex expression patterns. *Gene Expr Patterns* 2006, *6*, 581-588.

[10] Al-Sadi, R., Khatib, K., Guo, S., Ye, D., et al., Occludin regulates macromolecule flux across the intestinal epithelial tight junction barrier. *Am J Physiol Gastrointest Liver Physiol* 2011, *300*, G1054-1064.

[11] Krug, S. M., Amasheh, S., Richter, J. F., Milatz, S., et al., Tricellulin forms a barrier to macromolecules in tricellular tight junctions without affecting ion permeability. *Mol Biol Cell* 2009, *20*, 3713-3724.

[12] Hirabayashi, S., Tajima, M., Yao, I., Nishimura, W., et al., JAM4, a junctional cell adhesion molecule interacting with a tight junction protein, MAGI-1. *Mol Cell Biol* 2003, *23*, 4267-4282.

[13] Laukoetter, M. G., Nava, P., Lee, W. Y., Severson, E. A., et al., JAM-A regulates permeability and inflammation in the intestine in vivo. *J Exp Med* 2007, *204*, 3067-3076.

[14] Cohen, C. J., Shieh, J. T., Pickles, R. J., Okegawa, T., et al., The coxsackievirus and adenovirus receptor is a transmembrane component of the tight junction. *Proc Natl Acad Sci U S A* 2001, *98*, 15191-15196.

[15] Tsukita, S., Katsuno, T., Yamazaki, Y., Umeda, K., Tamura, A., Roles of ZO-1 and ZO-2 in establishment of the belt-like adherens and tight junctions with paracellular permselective barrier function. *Ann N Y Acad Sci* 2009, *1165*, 44-52.

[16] Umeda, K., Matsui, T., Nakayama, M., Furuse, K., et al., Establishment and characterization of cultured epithelial cells lacking expression of ZO-1. *J Biol Chem* 2004, *279*, 44785-44794.

[17] Noda, S., Tanabe, S., Suzuki, T., Differential effects of flavonoids on barrier integrity in human intestinal Caco-2 cells. *J Agric Food Chem* 2012, *60*, 4628-4633.

[18] Noda, S., Tanabe, S., Suzuki, T., Naringenin enhances intestinal barrier function through the expression and cytoskeletal association of tight junction proteins in Caco-2 cells. *Mol Nutr Food Res* 2013, *57*, 2019-2028.

[19] Honda, H., Pazin, M. J., Ji, H., Wernyj, R. P., Morin, P. J., Crucial roles of Sp1 and epigenetic modifications in the regulation of the CLDN4 promoter in ovarian cancer cells. *J Biol Chem* 2006, *281*, 21433-21444.

[20] Suzuki, T., Elias, B. C., Seth, A., Shen, L., et al., PKC eta regulates occludin phosphorylation and epithelial tight junction integrity. *Proc Natl Acad Sci U S A* 2009, *106*, 61-66.

[21] Sakakibara, A., Furuse, M., Saitou, M., Ando-Akatsuka, Y., Tsukita, S., Possible involvement of phosphorylation of occludin in tight junction formation. *J Cell Biol* 1997, *137*, 1393-1401.

[22] Raleigh, D. R., Boe, D. M., Yu, D., Weber, C. R., et al., Occludin S408 phosphorylation regulates tight junction protein interactions and. *J Cell Biol* 2011, *193*, 565-582.

[23] Jain, S., Suzuki, T., Seth, A., Samak, G., Rao, R., Protein kinase Czeta phosphorylates occludin and promotes assembly of epithelial tight junctions. *Biochem J* 2011, *437*, 289-299.

[24] Dorfel, M. J., Westphal, J. K., Huber, O., Differential phosphorylation of occludin and tricellulin by CK2 and CK1. *Ann N Y Acad Sci* 2009, *1165*, 69-73.

[25] McKenzie, J. A., Riento, K., Ridley, A. J., Casein kinase I epsilon associates with and phosphorylates the tight junction protein occludin. *FEBS Lett* 2006, *580*, 2388-2394.

[26] Seth, A., Sheth, P., Elias, B. C., Rao, R., Protein phosphatases 2A and 1 interact with occludin and negatively regulate the assembly of tight junctions in the CACO-2 cell monolayer. *J Biol Chem* 2007, *282*, 11487-11498.

[27] Sheth, P., Samak, G., Shull, J. A., Seth, A., Rao, R., Protein phosphatase 2A plays a role in hydrogen peroxide-induced disruption of tight junctions in Caco-2 cell monolayers. *Biochem J* 2009, *421*, 59-70.

[28] Dou, W., Zhang, J., Sun, A., Zhang, E., et al., Protective effect of naringenin against experimental colitis via suppression of Toll-like receptor 4/NF-kappaB signalling. *Br J Nutr* 2013, *110*, 599-608.

[29] Azuma, T., Shigeshiro, M., Kodama, M., Tanabe, S., Suzuki, T., Supplemental naringenin prevents intestinal barrier defects and inflammation in colitic mice. *J Nutr* 2013, *143*, 827-834.

[30] Al-Rejaie, S. S., Abuohashish, H. M., Al-Enazi, M. M., Al-Assaf, A. H., et al., Protective effect of naringenin on acetic acid-induced ulcerative colitis in rats. *World J Gastroenterol* 2013, *19*, 5633-5644.

In: Apigenin and Naringenin
Editor: Nilus M. Stacks

ISBN: 978-1-63463-987-3
© 2015 Nova Science Publishers, Inc.

Chapter 9

DIETARY SOURCES, BIOAVAILABILITY AND BIOLOGICAL ACTIVITIES OF NARINGENIN AND ITS DERIVATIVES

***Ramesh Prasad Pandey, Rit Bahadur Gurung and Jae Kyung Sohng*[*]**
Institute of Biomolecule Reconstruction, Department of Pharmaceutical Engineering, Sun Moon University, Republic of Korea

ABSTRACT

Naringenin is one of the most studied flavonoids abundant in citrus plants. Most of the flavonoids derived from plants are found to be highly beneficial to human health because of their intrinsic antioxidant potential. Recent research on these small molecules have expanded on their effects and the molecular mechanisms for their modes of action in the prevention of a wide range of diseases. Particularly, naringenin and their derivatives are known to exhibit strong anti-oxidant potential along with plenty of protective effects for the improvement of human health. Several *in vitro* and *in vivo* experimental results support their beneficial effects, however, the lack of adequate clinical studies have limited their use as therapeutics. In this chapter, the occurance and dietary sources of naringenin will be briefly discussed along with naringenin biosynthesis pathway in plants. Moreover, the chapter will focus on discussing several protective effects of the compounds and their derivatives against cancer, obesity, diabetes, hypertension, hyperlipidemia, neuro-degenerative disorders, and other metabolic syndromes.

1. INTRODUCTION

Flavonoids are some of the most studied and ubiquitous secondary plant metabolites. The biosynthetic pathways of most flavonoids have been elucidated and their significance in

[*] Corresponding author: Prof. Jae Kyung Sohng; E-mail: sohng@sunmoon.ac.kr; Fax: 82-41-544-2919; Tel: 82-41-530-2246.

plants and human health have been extensively studied. Among more than 9,000 known structures of flavonoids, a relatively small number of flavonoids have been identified that exhibit potential health promoting effects in humans and animals. However, most of the flavonoids have been found to play different roles in plants such as maintaining physiology and biochemistry, growth and development, signaling and protection from various types of stress (Anderson and Markham 2006).

Structurally, flavonoid groups of compounds contain C_6-C_3-C_6 backbone skeletons derived from a phenylpropanoid (C_6-C_3) and three molecules of C_3 elongation units of malonyl-CoA (Anderson and Markham 2006; Grayer 1989). Based on the degree of oxidation and saturation in the heterocyclic C-ring, flavonoids are classified into various subgroups which includes flavan, flavan-3-ol, flavanone, dihydroflavonol, flavone, and flavonol (Winkel-Shirley 2006; Anderson and Markham 2006; Harborne 1993). Further alterations of the core unit of the flavonoid by structural rearrangements and enzymatic modifications such as hydroxylation, methylation, prenylation, alkylation, and glycosylation have diversified these small molecules to significantly higher numbers in the plant kingdom (Schwinn 2004; Grotewold 2006). Hydroxylation usually occurs at the C-3, C-5, C-7, and C-4' positions. Conjugation with sugars are very common, with the majority of flavonoids existing naturally as glycosides enhancing the water solubility of the compounds.

Naringenin, (2*S*)-5,7-dihydroxy-2-(4-hydroxyphenyl)-2,3-dihydro-4H-chromen-4-one, is a typical flavanone also known as (2*S*)-naringenin or 4',5,7-trihydroxyflavanone (Figure 1). Flavanones are non-planar and have a chiral center at C-2 attached with a B-ring in the α-configuration. Flavanones like naringenin and naringenin chalcone are inter-convertible under different pH conditions of the medium. For example, in alkaline conditions, naringenin is converted into chalcone and the reverse is true in acidic conditions. This chapter will focus on the occurance, biosynthesis, dietary sources, protective effects and health benifits of naringenin and its derivatives in detail.

1.1. Occurance of Narningenin

Naturally, naringenin is present in the 2*S* configuration. However, a change in the physical and chemical environment during the extraction and purification processes can cause racemisation of naringenin (Grayer 1989). Naringenin occurs in aglycone as well as glycoside forms in nature. However, other enzymatic modifications alter the structure of naringenin to various flavonoids as it is one of the major intermediates of the flavonoid biosynthetic pathway. Aglycones like naringenin are usually stored as glycosides in plants, conjugating various monosaccharides as well as disaccharides at different hydroxyl positions. 7-*O*-, and 4'-*O*-glucosides and glycosides of naringenin are abundant in comparison to 5-*O*-glycosides, which are very rare, since the 5-hydroxyl group forms hydrogen bonds with the adjacent 4-carbonyl group. A few C-glycosides of naringenin have also been reported, though the majority of naringenin derivatives are present as *O*-glycosides. The most common glycoside units attached to naringenin include D-glucose and L-rhamnose.

Glucose conjugated at the 7-hydroxyl position (naringenin 7-*O*-glucoside, prunin) or at the 4'-hydroxyl group (naringenin 4'-*O*-glucoside) of naringenin are commonly reported glucosides from various plants. A glucose unit conjugated at C-6 (naringenin 6-C-β-D-glucoside) was isolated from *Tulipa gesneriana* (Budzianowski and Skrzypczakowa 1978) whereas C-8 glucoside (naringenin 8-C-β-D-glucoside, isohemiphloin) was reported in *Bowdichia virgilioides* (Velozo et al., 1999). Both C-6 and C-8 glucose conjugated derivatives of naringenin have also been isolated from *Vigna umbellata*. Apart from the glucosides of naringenin, rutinosides (6-*O*-α-L-rhamnosyl-D-glucosides) and neohesperidosides (2-*O*-α-L-rhamnosyl-D-glucosides) attached at the 7-hydroxyl position, narirutin and naringin respectively, are abundant in citrus fruit. For example, narirutin is found in *Citrus sinensis*, *C. paradisi*, and *C. Reticulata*, whereas naringin is present in *C. aurantium* and *C. paradisi*. Similarly, naringenin 8-C-neohesperidoside is found in fruits of *Eriobotrya japonica* (loquat) (Yannai 2004). Naringenin 7-rutinosides are generally tasteless, but the neohesperidosides attached at the same position of naringenin have an intensely bitter taste. In contrast, the neohesperidoside of naringenin dihydrochalcone is used a hypocaloric sweetner and has been approved by the European Union and United States Food and Drug Administration (FDA) as a food additive (European Parliament and Council Directive 1994; Lindley 1996). The occurance of flavanone glucosides are comparatively rare in comparision to other flavonoid glucosides in most plants. Naringenin 7-*O*-glucuronoide was isolated from *Scutellaria baicalensis* Georgi, a traditional chinese herbal plant belonging to the Labiatae family (Liu et al., 2011). Besides citrus plants, naringenin aglycone is also found in *Cycliopia intermedia* and potatoes (Ferreira et al., 1998; Lewis et al., 1998). Prenylated naringenin derivatives are also present in these plants. C-6 and C-8 prenyl naringenin, and C-6 geranyl naringenin are found in hops, *Humulus lupulus* and hops-containing foods and drinks such as beer (Stevens et al., 1999; Crozier et al., 2009). 7-*O*-geranyl naringenin was also enzymatically synthesized by Kuzuyama and colleagues (Kuzuyama et al., 2005).

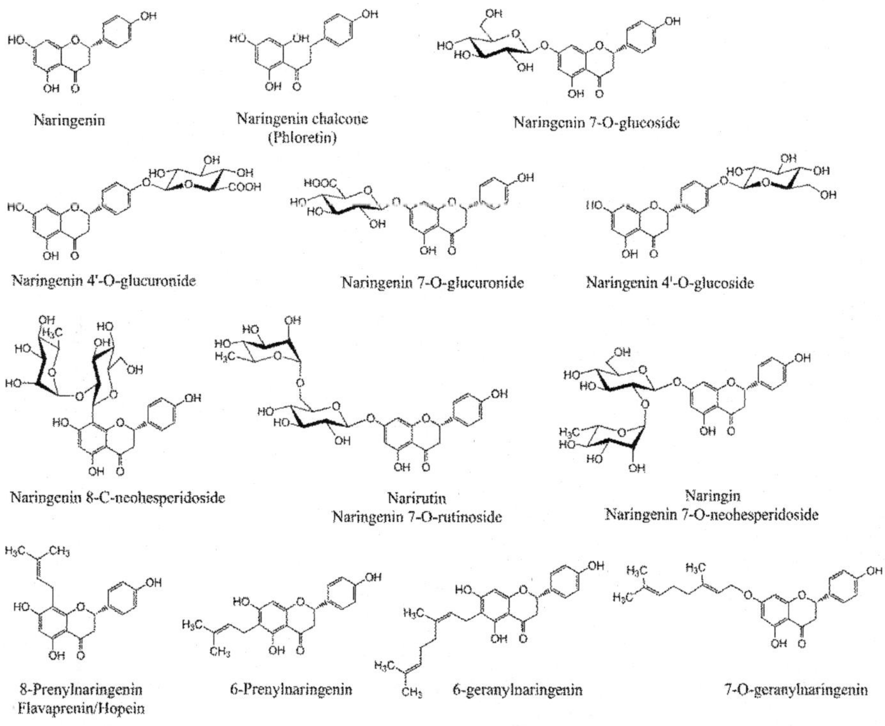

Figure 1. Structures of naringenin and its derivatives identified from various natural sources.

1.2. Biosynthesis of Naringenin

The flavonoid biosynthetic pathway is one of the most studied and well elucidated plant secondary metabolite pathways. The functions of the proteins responsible for the catalysis of each step of the flavonoid biosynthesis are well known and characterized. Naringenin is one of the intermediates of the flavonoids biosynthetic pathways. Naringenin chalcone is produced by the condensation of three molecules of malonyl-CoA generated via citrate, an intermediate product of tri-carboxalic acid (TCA) cycle and one molecule of 4-coumaroyl-CoA produced via the phenylpropanoid or aerogenate pathways. Acetyl-CoA, produced in the cytosol of the cell by ATP-citrate lyase, is utilized by acetyl-CoA carboxylase (ACC) to form malonyl-CoA. Another major precursor of the naringenin biosynthesis pathway, 4-coumaroyl-CoA, is produced via the phenylpropanoid pathway (Figure 2). The aromatic amino acids phenylalanine or tyrosine generated in the shikimate and aerogenate pathways undergo successive enzymatic conversions to produce 4-coumaroyl-CoA.

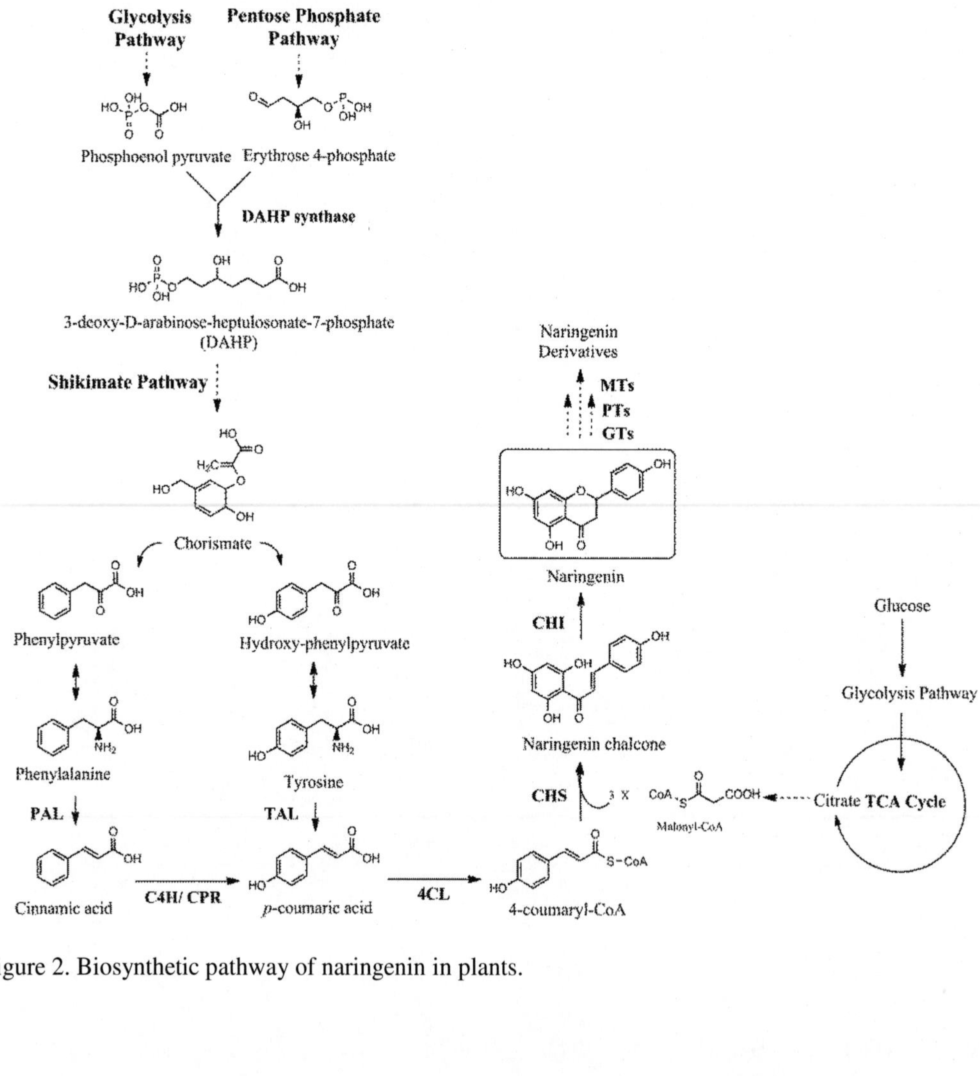

Figure 2. Biosynthetic pathway of naringenin in plants.

Generally, the biosynthesis of naringenin in plants involves five to six enzyme catalyzed reactions from tyrosine and phenylalanine (Winkel-Shirley 2001; Koopman et al., 2012; Gholami et al., 2014). Tyrosine ammonia lyase (TAL) catalyzes the conversion of tyrosine to *p*-coumaric acid in a single step, whereas phenylalanine conversion to *p*-coumaric acid requires two step modifications (Figure 2). Phenylalanine ammonia lyase (PAL) converts phenylalanine to *trans*-cinnamic acid through *trans*-elimination of ammonia along with the pro-3*S* proton, which is subsequently converted to *p*-coumaric acid by cinnamic acid 4-hydroxylases (C4H) and its associated cytochrome p450 reductase (CPR). The use of tyrosine as a precuror substrate for the biosynthesis of naringenin circumvents the use of membrane bound C4H and CPR proteins while using phenylalanine as an intermediate substrate (Wang et al., 2011; Santos et al., 2011). 4-coumaric acid-CoA ligase (*4CL*) converts *p*-coumaric acid to 4-coumaroyl-CoA for the condensation with malonyl-CoA catalyzed by chalcone synthase (*CHS*) to produce naringenin chalcone. Another final step enzyme, chalcone isomerase (*CHI*), converts the aforementioned naringenin chalcone into naringenin by forming heterocyclic C-ring (Lin et al., 2014; Fowler and Koffas 2009; Pandey and Sohng 2013). Thus, the naringenin produced is modified by various hydroxylase and cytochrome P450 enzymes to produce diverse kinds of flavonoids. The post modification enzymes like glycosyltransferases (GTs), methyltransferases (MTs), and prenyltransferases (PTs) also diversify naringenin into various derivatives.

1.3. Dietary Sources of Naringenin

Flavonoids are ubiquitous in all plants, including most of the edible fruits and vegetables. However, the type of flavonoid present in different dietary sources vary. The most consumed commodities around the globe that are rich in polyphenolic compounds are tea, wine, coffee and cocoa/chocolate. The polyphenols contents in vegetables and fruits are comparatively lower. However, some herbs and spices as well as nuts and algae are rich sources of polyphenolic compounds (Crozier et al., 2006). Flavanones appear exclusively in citrus plants and apples. Thus, citrus fruits and plant products (juices and tissues) are major dietary sources of naringenin. Infact, the distrubution of the molecules varies within the plant, with much higher concentrations in the solid tissues compared to the juice (Tomás-Barberán and Clifford 2000). Fresh tomatoes and potatoes as well as their products are also sources of naringenin and naringenin chalcone (Krause and Galensa 1992). Grape fruit juice contains significantly higher amounts of naringin, ranging from 295 to 377 mgl^{-1} (McIntosh and Mansell 1997). The peels of orange and grape fruits, including albedo and flavedo, are particularly rich sources of naringin (Wu et al., 1991). Additionally, narirutin is abundant in oranges and mandarins whereas naringenin, narirutin, and naringin are present in grape fruits (Kawaii et al., 1999; Harapu et al., 2010). Recently reported novel sources of naringenin and its glycoside derivatives are *Litchi chinensis* (Ma et al., 2014), berries harvested in Serbia (Pavlović et al., 2013), shells of *Bixa Orellana* L (Yousuf et al., 2013), *Withania somnifera* (ashwagandha) (Alam et al., 2011), and *Camellia sinensis* green tea seeds (Hasegawa et al., 2011).

1.4. Bioavailability of Naringenin and Its Glycosides

Most of the flavanone and its glycosides, including naringenin 7-*O*-rutinoside, ingested as fruit juice are absorbed in the intestine (Ameer et al., 1996; Erlund 2001). However, naringenin metabolites were not observed in the plasma, although glucuronide derivatives, naringenin 7-*O*-glucuronide, naringenin 4'-*O*-glucuronide, and naringenin-*O*-diglucuronide were identified in their urine samples, along with other flavanone metabolites (Manach et al., 2005; Mullen et al., 2008). When higher concentrations of hesperetin 7-*O*-rutinoside (168 μmol) and naringenin 7-*O*-rutinoside (12 μmol) were ingested, analysis of the plasma and urine samples recovered only 6.5% of hesperetin metabolites and 17.1% of naringenin metabolites compared to the amounts ingested. This result proves that naringenin 7-*O*-rutinoside is more bioavailable than its hesperetin counterpart. The study also illustrated that the 3' and 4' substituents play crucial roles in absorption of the molecules in the large intestine (Manach et al., 2003; Manach et al., 2005; Mullen et al., 2008; Crozier 2009). Similar results of higher bioavailability of naringenin-*O*-glycosides compared to other apigenin and hesperetin glucosides have been reported recently (Escudero-López et al., 2014).

Another study showed that the bioavailability of naringenin was increased by conversion from rutinoside to glucoside independent of absorption site and increased dose of naringenin 7-*O*-rutinoside (Bredsdorff et al., 2010). However, a study suggested that the bioavailability of flavanone is dependent on the occurrence of specific microbiota able to remove the sugar units like rutinosides from the glycosides after that the aglycone is absorbed from the gut (Vallejo et al., 2010).

Prenylation of naringenin and other flavonoids was found to alter their bioavailability and bioaccumulation. Studies showed that the prenylation of naringenin increased its accumulation in muscle tissue and restricted the efflux from hepatocytes to blood and enterohepatic circulations, resulting in slow excretion and rapid absorption of prenylated flavonoids (Terao and Mukai 2014).

Recently, the dissolution and bioavailability properties of naringenin were improved from poor water solubility and slow dissolution after oral ingestion by solid dispersion (Khan et al., 2014). In a separate recent study, Claudie and colleagues reported that citrus flavonoids like hesperetin, hesperidin and ascorbic acid along with naringenin enhances the bioavailability of other phytochemicals like carotenoids (Claudie et al., 2013). Similarly, naringenin was found to enhance the oral bioavailability of felodopine (Surya Sandeep et al., 2014).

The bioavailability of naringenin was enhanced by preparing a complex of naringenin with hydroxypropyl-*β*-cyclodextrin, which increased its water solubility by 400 folds and its transport across Caco-2 cells of the gut epithelium by 11-fold (Shulman et al., 2011). The cooking of cherry tomatoes also enhanced the bioavailability of naringenin in humans compared to fresh cherry tomatoes (Bugianesi et al., 2004). The absolute bioavailability of naringenin was comparatively very low (limited to 4%) and was increased to 8% when orally administered in combination with naringin in rabbits. Thus, oral dosing of naringin resulted in significantly higher bioavailability of naringenin.

Naringenin and naringin metabolites were detected in the form of glucuronides and sulfates (Hsiu et al., 2002). Another study showed that the intestinal absorption of naringenin and naringenin 7-*O*-glucoside was almost similar. However absorption of naringin-7-

rhamnoglucoside was lower, indicating the role of the glycosidic moiety in the bioavailability of naringenin (Felgines et al., 2000).

2. Biological Activities: Protective Effects and Health Benefits

Medicinal plants, plant fruits and their products are known to have potential health beneficial effects on humans and animals and have been employed for the treatment of various diseases since ancient times. Citrus plants, particularly the fruits of oranges, grapes, mandarins, lemons, and limes in addition to other citrus fruits are rich sources of flavonoids like naringenin and its derivatives which are known to possess various biological activities. Naringenin, its glycoside derivatives and other analogues were found to exhibit strong anti-inflammatory and antioxidant activities. They were also found to be beneficial for preventing the onset of chronic diseases like diabetes, obesity, hypertension, and cancers. A number of molecular mechanisms have also been proposed and elucidated to prove the potential activities of these compounds. However, sufficient clinical evidences are not available to use these compounds as therapeutic agents for the treatment of diseases. Only limited studies have been performed with detailed analysis of the pharmacokinetics of naringenin and its derivatives. Most of the studies were conducted in rat models and lack sufficient information on their *in vivo* activities. In this section, we will discuss recent advances in understanding the biological activities of naringenin and its derivatives.

2.1. Anti-Cancer Properties

Naringenin is reported to have anti-proliferative effects in many cancer cell lines. Antioxidant activities, protein kinase activity, and glucose uptake have been proposed as molecular mechanisms for these anticancer effects. There are a number of studies performed recently using naringenin and its derivatives to study their effects on various cancers either *in vitro* or *in vivo* in different animal models. For example, naringenin has been found to inhibit the migration of bladder cancer cells (TSGH-8301) by repressing the matrix metalloproteinase (MMP-2), inhibiting AKT activity (protein kinase B) and blocking the nuclear translocation of nuclear factor κ-light chain enhancer in activated B-cells (Liao, et al., 2014). Similarly, it has inhibitory activity on the growth of human hepatocellular carcinoma cells (HepG2 cells) via inhibition of cell proliferation and apoptosis induction through the activation of caspase-3 (Arul and Subramanian 2013). Moreover, its derivative 6-C-(E-phenylethenyl)-naringenin also suppresses colorectal cancer growth by inhibiting cyclooxygenase-1 (COX-1), which plays a critical role in human colorectal carcinogenesis (Li et al., 2014). Naringenin also exhibited antiproliferative and apoptotic effects by decreasing the expression of proliferating cell nuclear antigens and B-cell lymphoma 2 (Bcl-2) while increasing the expression of Bcl-2 associated X protein (Bax) and caspase-3 in *N*-nitrosodiethylamine (NDEA)-induced hepatocarcinogenic rats (Subramanian and Arul 2013).

Naringenin also helps in the prevention of colon cancer (Leonardi et al., 2010), glial tumor cell proliferation (Sabarinathan et al., 2011), and it decreases the invasiveness and metastatis of the tumor cells by inhibiting transforming growth factor (TGF)-*β*-induced epithelial-to-mesenchymal transition in pancreatic cancer cells (Lou et al., 2012).

Based on previous results, naringenin modified at the 7th position by bulky substituents have shown anti-carcenogenic effects against human colon cancer cells (HCT116). A novel derivative of naringenin modified at the 4th position also exhibited improved anti-cancer as well as anti-oxidative effects (Yoon et al., 2013). Naringenin derivatives modified at the 7th hydroxyl position have also been reported to exhibit significant apoptosis-inducing effects (Lee et al., 2008). Other synthetic naringenin derivatives like 5-hydroxy-7,4'-diacetyloxyflavanone-*N*-phenyl hydrazine were found to induce apoptosis via up-regulation of Fas/FasL expression and inhibition of phosphoinositide 3-kinase (PI3K)/Akt signaling pathways in non-small-cell lung cancer cell lines (Bak et al., 2011).

Recently, the combination of naringenin and tamoxifen at lower concentrations was found to be more effective in impairing cell proliferation and viability of breast cancer cells (MCF-7) than each compound independently (Hatkevich et al., 2014). Alpha-tocopherol and naringenin have shown synergistic effects in transglutaminase-induced differentiation of human prostate cancer cells (PC-3) (Torricelli et al., 2011). Similar synergistic effects of angioinhibitory and antitumor activity were observed when naringenin and curcumin were administered in Swiss albino mice intraperitoneally implanted with 1×10^6 Ehrlich ascites carcinoma cells (Anand et al., 2012). Naringenin also enhanced the anti-tumor effect of doxorubicin by selectively modulating drug efflux pathways. Thus, it could be a useful adjunct to improve the effectiveness of chemotherapeutic agents in treatment of human cancers (Zhang et al., 2009). Naringenin has also been suggested as an ideal surgical adjuvant therapy for breast cancer patients as it reduces lung metastasis by regulating host immunity (Qin et al., 2011). A study of naringenin-loaded nanoparticles against 7, 12-dimethylbenz(a)anthracene induced experimental oral carcinogenesis showed more potent anti-lipid peroxidative antiproliferative effects and antioxidant potential compared to free naringenin, proving its chemopreventive efficacy (Sulfikkarali et al., 2013).

2.2. Against Hyperglycemia and Diabetes

Increased blood sugar level and resistance to insulin result in diabetes. Generally, high-fat diets increase inflammatory cytokines (tumor necrosis factor- *α* (TNF-*α*), interleukin-6 (IL-6)), as it has been found that these cytokines were elevated in individuals with insulin resistance and type II diabetes (Pickup et al., 2000; Kado et al., 1999). High fat diets also cause insulin resistance and hyperglycemia (Terra et al., 2009; Lee et al., 2010). The effects of naringenin and its glycoside naringin on insulin resistance and diabetes in animal and cell studies have been well documented (Jung et al., 2004; Alam et al., 2013). Herein, we have noted a few studies in which naringenin and naringenin glycoside derivatives were found to have positive effects on reducing blood sugar level, thus preventing the development of fatal diseases like diabetes and other disorders. For example, naringenin-7-*O*-*β*-D-glucoside was found to lower blood glucose as well as total lipid level in streptozotocin induced diabetic rats (Choi et al., 1991). Since a high level of reactive oxygen species is produced in diabetes mellitus, which is detrimental for oxidative damage of deoxyribonucleic acid (DNA), a study of naringenin in alloxan-induced diabetic mice found that the compound showed DNA-protective effects, thus helping in the prevention of further deterioration from the disease

(Oršolić et al., 2011). In another recent experiment, it was observed that naringenin inhibits α-glucosidase activity, thus helping in the management of diabetes type II (Priscilla et al., 2014). Similarly, Annadurai and colleagues found that naringenin alleviates hyperglycemia-mediated inflammation in hepatic and pancreatic tissues in diabetes mellitus rats (Annadurai et al., 2013).

The mechanism for the improvement of hyperlipidemia and hyperglycemia by naringenin has been well studied in human hepatoma cells (HepG2), proving the inhibition of apolipoprotein B by activation of PI3K. This is independent of insulin receptor substrate-1 phosphorylation and explains the insulin-like effects of naringenin (Borradaile et al., 2003; Mulvihill et al., 2009). Naringenin was found to decrease the microsomal triglyceride transfer protein essential for the hepatocyte assembly and secretion of apolipoprotein B100. The study also showed that naringenin could activate pathways similar to insulin for regulation of hepatocyte lipid homeostasis (Allister et al., 2005).

2.3. Neuroprotective Activities

The structural damage of neurons and their supporting structures is mainly caused by cerebral pathogenesis as a result of oxidative stress and inflammation. Many natural products, particularly polyphenols, neutralize the reactive oxygen species spontaneously generated in the body by metabolism processes and minimize the chance of damage by oxidative stress. The potential neuroprotective role of naringenin and the underlying mechanisms has been studied by both *in vitro* as well as *in vivo* experiments.

A study has reported the potential of naringenin to reduce oxidative stress and maintain mitochondrial membrane integrity. Meanwhile, it was also found that naringenin downregulated pro-apoptotic genes (Bax and Caspase-3) while upregulating anti-apototic gene (Bcl2) in carbaryl induced mouse neuroblastoma cells (Neuro 2A) (Muthaiah et al., 2013). Similarly, naringenin was found to protect from ischemic brain injury by suppressing nuclear factor ƙ-light-chain-enhancer of activated B cells (NF-ƙB) protein mediated neuroinflammation, serving as a potential neuroprotectant in patients at high risk of ischemic stroke (Raza et al., 2013). A separate study on focal cerebral ischemia-induced male Sprague-Dawley rats followed by administration of naringenin showed the down regulation of nucleotide-binding oligomerization domain-containing protein 2 (NOD2), receptor-interacting serine/threonine-protein kinase 2 (RIP2), and NF-ƙB in the ischemic cerebral cortex. Additionally, the activity of claudin-5 was found to be up-regulated, indicating the neuroprotective effect of naringenin (Bai et al., 2014). Likewise, it protects retinal neurons against structural and functional damage caused by *N*-methyl-*N*-nitrosourea, suggesting the possible application of naringenin for topical eye treatment (Lin et al., 2014). Recent studies found that severe abnormalities in brain glucose/energy metabolism and insulin signaling play crucial roles in early sporadic Alzheimer's disease pathology. The study conducted to identify the protective effect of naringenin on dementia induced by an intracerebroventricular (ICV) injection of streptozotocin (STZ) showed that naringenin treatment had positive effects on recovery from Alzheimer's disease in the early stages. The naringenin treated rats had better learning and memory performance than rats treated with saline (Yang et al., 2014). In addition to Alzheimer's disease, it also protects against oxidative stress (neurotoxicity) induced in Parkinson's disease by enhancing nuclear factor E2-related factor 2 (Nrf2) protein levels and subsequent activation of antioxidant response element (ARE) pathway genes in SH-SY5Y cells and in mice (Lou et al., 2014). Iron plays a crucial role in the human body as it is one of the components of metalloproteins involved in metabolic processes like oxygen sensing and transport as well as energy metabolism. During pathogenesis of Alzheimer's and Parkinson's diseases, iron accumulates gradually. A recent study found that the protective effects of naringenin on iron induced cerebral neurotoxicity correlated with oxidative stress (Chtourou et al., 2014). Naringenin has also been found to work as an anti-depressant by activating brain-derived neurotrophic factor (BDNF) signaling (Yi et al., 2014). These experimental evidences clarify that naringenin and a few of its derivatives have the potential to protect the nervous system by preventing the functional or structural damage of neurons or parts of the nervous system.

2.4. Anti-Inflammatory Activities

Like other polyphenols and flavonoids, naringenin was also found to exhibit anti-inflammatory effects, protecting various human and animal parts and organs from inflammation induced diseases/pathogenesis. Anti-inflammatory effects of naringenin and other phenolic compounds isolated from various plants have been broadly studied (Hsu et al., 2013; Lim et al., 2013; Coelho et al., 2013). Naringenin has been found to protect the liver from oxidative stress by activating the nuclear translocation of nuclear factor erythroid 2 (NF-E2)-related factor 2 and attenuating TNF-α pathway, exhibiting anti-inflammatory activity in liver tissue (Esmaeili and Alilou 2014). Similarly, it has exhibited analgesic effects by alleviating neuropathic pain by inhibiting neuroinflammation in rats (Hu and Zhao 2014). Naringenin suppresses the development of atopic dermatitis-like skin lesions by reducing interferon-γ production of activated CD4 (+) T cells (Kim et al., 2013). Interestingly, naringenin was found to attenuate hepatic macrophage infiltration and inflammation stimulated by dietary cholesterol, protecting from atherosclerosis (Assini et al., 2013). It also inhibits inflammatory signaling in glial cells and protects against neuroinflammatory injury (Vafeiadou et al., 2009). Naringin was also found to exhibit antitussive, anti-airway hyper responsiveness and anti-inflammation effects on chronic cigarette smoke exposure-induced chronic bronchitis in guinea pigs, possessing novel therapeutic potential in the treatment of chronic bronchitis (Luo et al., 2012) and allergen-induced airway inflammation in asthma induced murine model (Shi et al., 2009). It protects from experimental colitis by suppressing Toll-like receptor 4/NF-kB signaling (Dou et al., 2013).

Chlamydia trachomatis, a sexually transmitted bacterial infection, can manifest itself as either acute cervicitis, pelvic inflammatory disease, or a chronic asymptomatic infection. Inflammation induced by *C. trachomatis* greatly contributes to the pathogenesis of the disease. The flavonoid naringenin down regulated the inflammatory cytokines (GM-CSF, TNF, IL-1β, IL-1α, IL-6, IL-12p70, and IL-10) and chemokines (C-C motif ligand (CCL)4, CCL5, C-X-C motif ligand (CXCL)1, CXCL5, and CXCL10) in a dose dependent manner, exhibiting its ability to modulate inflammatory mediators produced as a result of *C. trachomatis* (Yilma et al., 2013).

2.5. Against Obesity

Naringenin and other citrus polyphenols have suppressed body weight gain, fat accumulation, and development of hyperlipidemia and hyperglycemia by up regulating transcriptional expression of enzymes which are involved in the β-oxidation, for example, peroxisome proliferator activated receptor (PPAR) α, acyl-CoA oxidase, fatty acid synthase in liver and white adipose tissue in mice (Fukuchi et al., 2008). Similarly, naringenin has lowered total triglyceride and cholesterol contents in plasma and liver in rats whereas adiposity and triglyceride contents were also lowered in parametrial adipose tissue by up regulating the expression of PPARα, carnitine-palmitoyl transferase 1L (CPT-1), and uncoupling protein 2 (UCP2) (Cho et al., 2011). Naringenin was also found to inhibit adipocyte differentiation, as adipose tissue development and function are correlated with obesity, inhibiting adipogenesis by inhibiting insulin stimulated glucose uptake in a dose dependent fashion and impairing mature fat cell function. In addition, naringenin also inhibited adiponectin protein expression in mature murine and human adipocytes (Richard et al., 2013).

2.6. Against Hyperlipidemia and Hypertension

Hyperlipidemia is one of the metabolic disorders in which the plasma lipid concentration is above normal concentrations. Plant secondary metabolites, particularly flavonoids, are capable of lowering plasma lipid concentrations (Cook and Samman 1996; Bravo 1998). The supplementation of naringin in high-fat-diet-fed and high-cholesterol-diet-fed rats showed decreased levels of plasma lipid and cholesterol, respectively (Pu et al., 2012; Shin et al., 1999). Similar evidences demonstrating the effect of naringenin and other analogues on reducing the lipid level in the body have been reported (Li et al., 2004; Mulvihill et al., 2009; Alam et al., 2013). In a clinical trial, naringenin was found to reduce the plasma total-cholesterol and low density lipid cholesterol concentrations in hypercholesterolemic patients (Jung et al., 2003).

Hypertension is a chronic medical condition in which elevated blood pressure is recorded in patients. The blood pressure is increased because of either high carbohydrate or fat level in the blood. Naringin has been found to improve hypertension in high-carbohydrate and high-fat-diet fed rats (Alam et al., 2013), along with stroke-prone hypertensive rats (Ikemura et al., 2012) by vascular relaxation. Naringenin also inhibited the TNF-α induced vascular smooth muscle cells proliferation and migration, which is one of the critical events in the pathogenesis of hypertension (Chen et al., 2012).

2.7. Cardioprotective Effects

Flavonoids exhibit beneficial health effects on cardiovascular functions because of their endogenous antioxidant properties. Besides exhibiting a number of health protective effects, they are also found to exert cardioprotective effects against myocardial ischaemia/reperfusion (I/R) injury (Cook et al., 1996; Qin et al., 2008). In a recent study in Langendorff-perfused rat hearts, naringenin and other flavanones substituted at the 5-hydroxyl position conferred an improved post-ischaemic functional recovery associated with lower extension of tissue injury (Testai et al., 2013). Naringenin, along with other flavonoids, have exhibited cardioprotective effects in a model of spontaneously beating, cultured adult rat cardiomyocytes against daunorubicin-induced toxicity (Mojzisová et al., 2006). Similar experiments showing the protective effects of naringenin on daunorubicin-induced toxicity in H9c2 cardiomyoblasts have been noted at significantly higher concentrations of naringenin and quercetin (Mojzisová et al., 2009). Naringin supplementation has prevented isoproterenol-induced myocardial infarction in rats. It has also reduced lipid peroxidation and inflammatory fibrosis, and improved antioxidant enzymes (Rajadurai et al., 2006). Similarly, it was found that naringenin prevents mitochondrial dysfunction during isoproterenol-induced myocardial infarction in male albino Wistar rats (Rajadurai et al., 2007). Naringin has been found to protect from doxorubicin induced cardiotoxicity without altering the antineoplastic activity of doxorubicin (Reddy et al., 2008).

The mechanism of protection of cardiomyocytes by naringin has been reported recently. Naringin protects cardiomyocytes via high-glucose and reactive oxygen species induced mitogen-activated protein kinase (MAPK) signal mediated pathways. In addition, it was found to prevent the high-glucose induced phosphorylation of p38 MAPK, extracellular signal-regulated kinase (ERK)1/2, and c-Jun N-terminal kinase (JNK) and also reduced apoptosis in H9c2 cardiac cells (Chen et al., 2014). It has been found that naringin decreased the mitochondrial Bax and Bcl-2 homologous antagonist/killer (Bak) expression while Bcl-2 expression was increased in H9c2 cells. It also modulated high-glucose-induced apoptosis followed by inactivation of caspase-3,-8, and -9 (Huang et al., 2013).

2.8. Anti-Atherosclerosis

The invasion and accumulation of both dead and living cells along with cholesterol, triglycerides, and crystalized materials leads to the formation of atherosclerosis, which is also known as arteriosclerotic vascular disease. The eventual progression of atherosclerosis leads to increased blood pressure and upon further complications, results in myocardial infraction, also commonly known as stroke. The consumption of plant flavonoids reduces low density lipoprotein lipid peroxidation by scavenging reactive oxygen/nitrogen species, chelating transition metal ions, sparing low density lipoprotein-associated antioxidants, macrophage oxidative stress and ultimately protecting lipid peroxidation in arterial cells, attenuating the development of atherosclerosis (Fuhrman and Aviram 2001). In a recent report, naringenin was found to decrease the progression of atherosclerosis by improving dyslipidemia in Western-diet-fed low-density lipoprotein receptor-null mice, leading to decreased atherosclerosis. Thus, it is suggested that the supplementation of citrus flavanones like naringenin could be a potential therapeutic strategy for hyperlipidemia and increased risk of atherosclerosis associated with insulin resistance (Mulvihill et al., 2010). Similarly, the anti-atherogenic activity of naringenin and naringin has been observed in high cholesterol-fed rabbits since there was decreased hepatic acyl-CoA:cholesterol acyltransferase (ACAT) activity, vascular cell adhesion molecule-1 (VCAM-1) and monocyte chemotactic protein-1 (MCP-1) expression levels in flavonoids-fed animal models (Lee et al., 2001). In the same hypercholesterolemic rabbit models, the feeding of naringin effectively reduced the fatty streak area in the thoracic aorta on macroscopic analysis and significantly reduced sub intimal foam cell infiltration on microscopic morphometry. It also inhibited hypercholesterolemia-induced intercellular adhesion molecule-1 (ICAM-1) expression on endothelial cells, thus exhibiting antiatherogenic effect (Choe et al., 2001).

CONCLUSION

Flavonoids are one of the most potent plant secondary metabolites for the treatment of various kinds of chronic diseases. Though vitamin C and vitamin E are currently considered the most effective antioxidants for preventing oxidative damage of biological parts and systems (Haramaki et al., 1998; Nagel et al., 1997), recent studies have uncovered severe side effects upon treatment of diabetes patients (Lonn et al., 2005; GISSI-Prevenzione Investigators 1999, Brown and Crowley 2005, Sesso et al., 2008). Hence, researchers are exploring the applications of flavonoids against a wide range of diseases like cancer, diabetes, hypertension, hyperlipidemia, atherosclerosis, inflammation, neuronal disorders, and cardio-failures. Particularly, citrus flavanones, namely naringenin, its derivatives including natural glycosides and unnatural synthesized analogues have been studied in detail for their possible application in the treatment of various diseases. The *in vitro* and *in vivo* studies of these compounds using various model animals have proved them as possible lead compounds for development as therapeutic agents for the prevention of cancers, diabetes, inflammation, and other neurodegenerative diseases. The possible mechanisms of action and their effects upon consumption or supplementation have also been studied. However, these evidences are not sufficient to prescribe these compounds as drugs. Hence, further clinical evidences and trials are necessary to determine the exact health beneficial effects of these compounds and ultimate use as drugs in the future. Nevertheless, the consumption of foods and drinks which are rich sources of these compounds on a daily basis in limited doses could certainly have positive impacts on the body.

REFERENCES

[1] Alam, MA; Kauter, K; Brown, L. Naringin improves diet-induced cardiovascular dysfunction and obesity in high carbohydrate, high fat diet-fed rats. *Nutrients,* 2013 5, 637-650.

[2] Alam, N; Hossain, M; Khalil, MI; Moniruzzaman, M; Sulaiman, SA; Gan, SH. High catechin concentrations detected in *Withania somnifera* (ashwagandha) by high performance liquid chromatography analysis. *BMC Complementary and Alternative Medicine*, 2011 11, 65.

[3] Allister, EM; Borradaile, NM; Edwards, JY; Huff, MW. Inhibition of microsomal triglyceride transfer protein expression and apolipoprotein B100 secretion by the citrus flavonoid naringenin and by insulin involves activation of the mitogen-activated protein kinase pathway in hepatocytes. *Diabetes,* 2005 54, 1676-1683.

[4] Ameer, B; Weintraub, RA; Johnson, JV; Yost, RA; Rouseff, RL. Flavanone absorption after naringin, hesperidin, and citrus administration. *Clinical Pharmacology and Therapeutics,* 1996 60, 34-40.

[5] Anand, K; Sarkar, A; Kumar, A; Ambasta, RK; Kumar, P. Combinatorial antitumor effect of naringenin and curcumin elicit angioinhibitory activities *in vivo*. *Nutrition and Cancer,* 2012 64, 714-724.

[6] Anderson, OM; Markham, KR. Flavonoids chemistry, biochemistry and applications. *Boca Raton, FL: CRC Press;* 2006.

[7] Annadurai, T; Thomas, PA; Geraldine, P. Ameliorative effect of naringenin on hyperglycemia-mediated inflammation in hepatic and pancreatic tissues of Wistar rats with streptozotocin- nicotinamide-induced experimental diabetes mellitus. *Free Radiccal Research,* 2013 47, 793-803.

[8] Arul, D; Subramanian, P. Naringenin (citrus flavonone) induces growth inhibition, cell cycle arrest and apoptosis in human hepatocellular carcinoma cells. *Pathology Oncology Research,* 2013 19, 763-770.

[9] Assini, JM; Mulvihill, EE; Sutherland, BG; Telford, DE; Sawyez, CG; Felder, SL; Chhoker, S; Edwards, JY; Gros, R; Huff, MW. Naringenin prevents cholesterol-induced systemic inflammation, metabolic dysregulation, and atherosclerosis in Ldlr^{-}/$^{-}$ mice. *Journal of Lipid Research,* 2013 54, 711-724.

[10] Bai, X; Zhang, X; Chen, L; Zhang, J; Zhang, L; Zhao, X; Zhao, T; Zhao, Y. Protective effect of naringenin in experimental ischemic stroke: down-regulated NOD2, RIP2, NF-κB, MMP-9 and up-regulated claudin-5 expression. *Neurochemical Research,* 2014 39, 1405-1415.

[11] Bak, Y; Kim, H; Kang, JW; Lee, DH; Kim, MS; Park, YS; Kim, JH; Jung, KY; Lim, Y; Hong, J; Yoon, DY. A synthetic naringenin derivative, 5-hydroxy-7,4'-diacetyloxyflavanone-N-phenyl hydrazone (N101-43), induces apoptosis through up-regulation of Fas/FasL expression and inhibition of PI3K/Akt signaling pathways in non-small-cell lung cancer cells. *Journal of Agricultural and Food Chemistry*, 2011 59, 10286-10297.

[12] Borradaile, NM; de Dreu, LE; Huff, MW. Inhibition of net HepG2 cell apolipoprotein B secretion by the citrus flavonoid naringenin involves activation of phosphatidylinositol 3-kinase, independent of insulin receptor substrate-1 phosphorylation. *Diabetes,* 2003 52, 2554-2561.

[13] Bravo, L. Polyphenols: chemistry, dietary sources, metabolism, and nutritional significance. *Nutrition Reviews,* 1998 56, 317-333.

[14] Bredsdorff, L; Nielsen, IL; Rasmussen, SE; Cornett, C; Barron, D; Bouisset, F; Offord, E; Williamson, G. Absorption, conjugation and excretion of the flavanones, naringenin and hesperetin from alpha-rhamnosidase-treated orange juice in human subjects. *The British Journal of Nutrition,* 2010 103, 1602-1609.

[15] Brown, BG; Crowley, J. Is there any hope for vitamin E? *The Journal of the American Medical Association,* 2005 293,1387-1390.

[16] Budzianowski, J; Skrzypczakowa, L. 6-C-Glucosylnaringenin from *Tulipa gesneriana. Phytochemistry,* 1978 17, 2044.

[17] Bugianesi, R; Salucci, M; Leonardi, C; Ferracane, R; Catasta, G; Azzini, E; Maiani, G. Effect of domestic cooking on human bioavailability of naringenin, chlorogenic acid, lycopene and beta-carotene in cherry tomatoes. *European Journal of Nutrition.* 2004 43, 360-366.

[18] Chen, J; Guo, R; Yan, H; Tian, L; You, Q; Li, S; Huang, R; Wu, K. Naringin inhibits ROS-activated MAPK pathway in high glucose-induced injuries in H9c2 cardiac cells. *Basic and Clinical Pharmacology and Toxicology,* 2014 114, 293-304.

[19] Chen, S; Ding, Y; Tao, W; Zhang, W; Liang, T; Liu, C. Naringenin inhibits TNF-α induced VSMC proliferation and migration via induction of HO-1. *Food and Chemical Toxicology,* 2012 50, 3025-3031.

[20] Cho, KW; Kim, YO; Andrade, JE; Burgess, JR; Kim, YC. Dietary naringenin increases hepatic peroxisome proliferators-activated receptor alpha protein expression and decreases plasma triglyceride and adiposity in rats. *European Journal of Nutrition,* 2011 50, 81-88.

[21] Choe, SC; Kim, HS; Jeong, TS; Bok, SH; Park, YB. Naringin has an antiatherogenic effect with the inhibition of intercellular adhesion molecule-1 in hypercholesterolemic rabbits. *Journal of Cardiovascular Pharmacology,* 2001 38, 947-955.

[22] Choi, JS; Yokozawa, T; Oura, H. Improvement of hyperglycemia and hyperlipemia in streptozotocin diabetic rats by a methanolic extract of *Prunus davidiana* stems and its main component, prunin. *Planta Medica,* 1991 57, 208-211.

[23] Chtourou, Y; Fetoui, H; Gdoura, R. Protective effects of naringenin on iron-overload-induced cerebral cortex neurotoxicity correlated with oxidative stress. *Biological Trace Element Research,* 2014 158, 376-383.

[24] Claudie, DM; Alexandrine, D; Bertrand, C; Franck, T; Marie-Josephe, A. Citrus flavanones enhance carotenoid uptake by intestinal Caco-2 cells. *Food and Function,* 2013 4, 1625-1631.

[25] Coelho, RC; Hermsdorff, HH; Bressan, J. Anti-inflammatory properties of orange juice: possible favorable molecular and metabolic effects. *Plant Foods for Human Nutrition,* 2013 68, 1-10.

[26] Cook, NC; Samman, S. Flavonoids-chemistry, metabolism, cardioprotective effects, and dietary sources. *The Journal of Nutritional Biochemistry,* 1996 7, 66-76.

[27] Crozier, A; Clifford, MN; Ashihara, H. Plant secondary metabolites. Occurance, structure and role in the human diet. *Oxford: Blackwell Publishing,* 2006.

[28] Crozier, A; Jaganath, IB; Clifford, MN. Dietary phenolics:chemistry, bioavailability and effects on health. *Natural Product Reports,* 2009 26,1001-1043.

[29] Dou, W; Zhang, J; Sun, A; Zhang, E; Ding, L; Mukherjee, S; Wei, X; Chou, G; Wang, ZT; Mani, S. Protective effect of naringenin against experimental colitis via suppression of Toll-like receptor 4/NF-κB signalling. *The British Journal of Nutrition,* 2013 110, 599-608.

[30] Erlund, I; Meririnne, E; Alfthan, G; Aro, A. Plasma kinetics and urinary excretion of the flavanones naringenin and hesperetin in humans after ingestion of orange juice and grapefruit juice. *The Journal of Nutrition,* 2001 131, 235-241.

[31] Escudero-López, B; Calani, L; Fernández-Pachónm MS; Ortegam A; Brighenti, F; Crozier, A; Del Rio, D. Absorption, metabolism, and excretion of fermented orange juice (poly)phenols in rats. *Biofactors.* 2014 40, 327-335.

[32] Esmaeili, MA; Alilou, M. Naringenin attenuates CCl4 -induced hepatic inflammation by the activation of an Nrf2-mediated pathway in rats. Clinical and Experimental Pharmacology and Physiology, 2014 41, 416-422.

[33] European Parliament and Council Directive 94/35/EC. Sweeteners for use in foodstuffs. *Official Journal of the European Communities* 1994 237, 3-12.

[34] Felgines, C; Texier, O; Morand, C; Manach, C; Scalbert, A; Régerat, F; Rémésy, C. Bioavailability of the flavanone naringenin and its glycosides in rats. American Journal of Physiology. *Gastrointestinal and Liver Physiology.* 2000 279, G1148-G1154.

[35] Ferreira, D; Kamara, BI; Brandt, EV; Joubert, E. Phenolic compounds from *Cyclopia intermedia* (Honeybush tea)-1. *Journal of Agriculture and Food Chemistry,* 1998 46, 3406-3410.

[36] Fowler, ZL; Koffas, MA. Biosynthesis and biotechnological production of flavanones: current state and perspectives. *Applied Microbiology and Biotechnology,* 2009 83, 799-808.

[37] Fuhrman, B; Aviram, M. Flavonoids protect LDL from oxidation and attenuate atherosclerosis. *Current Opinion in Lipidology,* 2001 12, 41-48.

[38] Fukuchi, Y; Hiramitsu, M; Okada, M; Hayashi, S; Nabeno, Y; Osawa, T; Naito, M. Lemon polyphenols suppress diet-induced obesity by up-regulation of mRNA levels of the enzymes involved in beta-oxidation in mouse white adipose tissue. *Journal of Clinical Biochemistry and Nutrition,* 2008 43, 201-209.

[39] Gholami, A; De Geyter, N; Pollier, J; Goormachtig, S; Goossens, A. Natural product biosynthesis in *Medicago* species. *Natural Product Reports,* 2014 31, 356-380.

[40] GISSI-Prevenzione Investigators (Gruppo Italiano per lo Studio della Sopravvivenza nell'Infarto miocardico). Dietary supplementation with n-3 polyunsaturated fatty acids and vitamin E after myocardial infarction: results of the GISSI-Prevenzione trial. *The Lancet,* 1999 354, 447-455.

[41] Grayer, RJ. Flavanoids. In: Dey PM; Harborne JB, editors. Plant Phenolics. London: *Academic Press Ltd;* 1989; pp 283-324.

[42] Grotewold, E. The science of flavonoids. New York*: Springer;* 2006.

[43] Haramaki, N; Stewart, DB; Aggarwal, S; Ikeda, H; Reznick, AZ; Packer, L. Networking antioxidants in the isolated rat heart are selectively depleted by ischemia-reperfusion. *Free Radical Biology and Medicine,* 1998 25, 329-339.

[44] Harapu, CD; Miron, A; Cuciureanu, M; Cuciureanu, R. Flavonoids-bioactive compounds in fruits juice. *Revista Medico-Chirurgicala a Societatii de Medici si Naturalisti din Iasi,* 2010 114, 1209-1214.

[45] Harborne, JB. The flavonoids: Advances in research since 1986. London: *Chapman-Hall,* 1993.

[46] Hasegawa, T; Akutsu, K; Kishi, Y; Nakamura, K. Constituents of the green tea seeds of *Camellia sinensis. Natural Product Communications,* 2011 6, 371-374.

[47] Hatkevich, T; Ramos, J; Santos-Sanchez, I; Patel, YM. A naringenin-tamoxifen combination impairs cell proliferation and survival of MCF-7 breast cancer cells. *Experimental Cell Research,* 2014 327, 331-339.

[48] Hsiu, SL; Huang, TY; Hou, YC; Chin, DH; Chao, PD. Comparison of metabolic pharmacokinetics of naringin and naringenin in rabbits. *Life Science,* 2002 70, 1481-1489.

[49] Hsu, CL; Fang, SC; Yen, GC. Anti-inflammatory effects of phenolic compounds isolated from the flowers of Nymphaea mexicana Zucc. *Food and Function,* 2013 4, 1216-1222.

[50] Hu, CY; Zhao, YT. Analgesic effects of naringenin in rats with spinal nerve ligation-induced neuropathic pain. *Biomedical Reports,* 2014 2, 569-573.

[51] Huang, H; Wu, K; You, Q; Huang, R; Li, S; Wu, K. Naringin inhibits high glucose-induced cardiomyocyte apoptosis by attenuating mitochondrial dysfunction and modulating the activation of the p38 signaling pathway. *International Journal of Molecular Medicine,* 2013 32, 396-402.

[52] Ikemura, M; Sasaki, Y; Giddings, JC; Yamamoto, J. Preventive effects of hesperidin, glucosyl hesperidin and naringin on hypertension and cerebral thrombosis in stroke-prone spontaneously hypertensive rats. *Phytotherapy Research,* 2012 26,1272-1277.

[53] Jung, UJ; Kim, HJ; Lee, JS; Lee, MK; Kim, HO; Park, EJ; Kim, HK; Jeong, TS; Choi, MS. Naringin supplementation lowers plasma lipids and enhances erythrocyte antioxidant enzyme activities in hypercholesterolemic subjects. *Clinical Nutrition,* 2003 22, 561-568.

[54] Jung, UJ; Lee, M-K; Jeong, K-S; Choi, M-S. The hypoglycemic effects of hesperidin and naringin are partly mediated by hepatic glucose-regulating enzymes in C57BL/KsJ-db/db Mice. *The Journal of Nutrition,* 2004 134, 2499-2503.

[55] Kado, S; Nagase, T; Nagata, N. Circulating levels of interleukin-6, its soluble receptor and interleukin-6/interleukin-6 receptor complexes in patients with type 2 diabetes mellitus. *Acta Diabetologica,* 1999 36, 67-72.

[56] Kawaii, S; Tomono, Y; Katase, E; Ogawa, K; Yano, M. Quantitation of flavonoid constituents in *Citrus* fruits. *Journal of Agricultural and Food Chemistry,* 1999 47, 3565-3571.

[57] Khan, AW; Kotta, S; Ansari, SH; Sharma, RK; Ali, J. Enhanced dissolution and bioavailability of grapefruit flavonoid naringenin by solid dispersion utilizing fourth generation carrier. *Drug Development and Industrial Pharmacy,* 2014 (ahead of print).

[58] Kim, TH; Kim, GD; Ahn, HJ; Cho, JJ; Park, YS; Park, CS. The inhibitory effect of naringenin on atopic dermatitis induced by DNFB in NC/Nga mice. *Life Science,* 2013 93, 516-524.

[59] Koopman, F; Beekwilder, J; Crimi, B; van Houwelingen, A; Hall, RD; Bosch, D; van Maris, AJ; Pronk, JT; Daran, JM. De novo production of the flavonoid naringenin in engineered *Saccharomyces cerevisiae. Microbial Cell Factories.* 2012 11,155.

[60] Krause, M; Galensa, R. Determination of naringenin and naringenin-chalcone in tomato skins by reversed phase HPLC after solid phase extraction. *Zeitschrift Fuer Lebensmittel-Untersuchung Und Forschung,* 1992 1, 29-32.

[61] Kuzuyama, T; Noel, JP; Richard, SB. Structural basis for the promiscuous biosynthetic prenylation of aromatic natural products. *Nature,* 2005 435, 983-987.

[62] Lee, C-H; Jeong, T-S; Choi, Y-K; Hyun, B-H; Oh, G-T; Kim, E-H; Kim, J-R; Han, J-I; Bok, S-H. Anti-atherogenic effect of citrus flavonoids, naringin and naringenin, associated with hepatic ACAT and aortic VCAM-1 and MCP-1 in high cholesterol-fed rabbits. *Biochemical and Biophysical Research Communications*, 2001 284, 681-688.

[63] Lee, ER; Kang, YJ; Kim, HJ; Choi, HY; Kang, GH; Kim, JH; Kim, BW; Jeong, HS; Park, YS; Cho, SG. Regulation of apoptosis by modified naringenin derivatives in human colorectal carcinoma RKO cells. *Journal of Cellular Biochemistry,* 2008 104, 259-273.

[64] Lee, IS; Shin, G; Choue, R. Shifts in diet from high fat to high carbohydrate improved levels of adipokines and pro-inflammatory cytokines in mice fed a high-fat diet. *Endocrine Journal,* 2010 57, 39-50.

[65] Leonardi, T; Vanamala, J; Taddeo, SS; Davidson, LA; Murphy, ME; Patil, BS; Wang, N; Carroll, RJ; Chapkin, RS; Lupton, JR; Turner, ND. Apigenin and naringenin suppress colon carcinogenesis through the aberrant crypt stage in azoxymethane-treated rats. *Experimental Biology in Medicine (Maywood),* 2010 235, 710-717.

[66] Lewis, CE; Walker, JRL; Lancaster, JE; Sutton, KH. Determination of anthocyanins, flavonoids and phenolic acids in potatoes. 1: Coloured cultivars of *Solanum tuberosum* L. *Journal of the Science of Food and Agriculture,* 1998 77, 45-57.

[67] Li, H; Zhu, F; Chen, H; Cheng, KW; Zykova, T; Oi, N; Lubet, RA; Bode, AM; Wang, M; Dong, Z. 6-C-(E-phenylethenyl)-naringenin suppresses colorectal cancer growth by inhibiting cyclooxygenase-1. *Cancer Research,* 2014 74, 243-252.

[68] Li, WL; Zheng, HC; Bukuru, J; De Kimpe, N. Natural medicines used in the traditional Chinese medical system for therapy of diabetes mellitus. *Journal of Ethnopharmacology,* 2004 92, 1-21.

[69] Liao, AC; Kuo, CC; Huang, YC; Yeh, CW; Hseu, YC; Liu, JY; Hsu, LS. Naringenin inhibits migration of bladder cancer cells through downregulation of AKT and MMP-2. *Molecular Medicine Reports,* 2014 10, 1531-1536.

[70] Lim, R; Barker, G; Wall, CA; Lappas, M. Dietary phytophenols curcumin, naringenin and apigenin reduce infection-induced inflammatory and contractile pathways in human placenta, foetal membranes and myometrium. *Molecular Human Reproduction,* 2013 19, 451-462.

[71] Lin, JL; Wang, YD; Ma, Y; Zhong, CM; Zhu, MR; Chen, WP; Lin, BQ. Protective effects of naringenin eye drops on *N*-methyl-*N*-nitrosourea-induced photoreceptor cell death in rats. *International Journal of Ophthalmology,* 2014 7, 391-396.

[72] Lin, Y; Jain, R; Yan, Y. Microbial production of antioxidant food ingredients via metabolic engineering. *Current Opinion in Biotechnology,* 2014 26, 71-78.

[73] Lindley, MG. NHDC: recent findings and technical advances. In: Grenby, TH, editor. Advances in sweeteners. London: *Blackie Academic and Professional;* 1996; pp 240-252.

[74] Liu, G; Rajesh, N; Wang, X; Zhang, M; Wu, Q; Li, S; Chen, B; Yao, S. Identification of flavonoids in the stems and leaves of *Scutellaria baicalensis Georgi. Journal of Chromatography B,* 2011 879, 1023-1028.

[75] Lonn, E; Bosch, J; Yusuf, S; Sheridan, P; Pogue, J; Arnold, JM; Ross, C; Arnold, A; Sleight, P; Probstfield, J; Dagenais, GR; The HOPE and HOPE-TOO Trial Investigators. Effects of long-term vitamin E supplementation on cardiovascular events and cancer: a randomized controlled trial. *The Journal of the American Medical Association*, 2005 293,1338-1347.

[76] Lou, C; Zhang, F; Yang, M; Zhao, J; Zeng, W; Fang, X; Zhang, Y; Zhang, C; Liang, W. Naringenin decreases invasiveness and metastasis by inhibiting TGF-β-induced epithelial to mesenchymal transition in pancreatic cancer cells. *PLoS One,* 2012 7, e50956.

[77] Lou, H; Jing, X; Wei, X; Shi, H; Ren, D; Zhang, X. Naringenin protects against 6-OHDA-induced neurotoxicity via activation of the Nrf2/ARE signaling pathway. *Neuropharmacology*, 2014 79, 380-388.

[78] Luo, YL; Zhang, CC; Li, PB; Nie, YC; Wu, H; Shen, JG; Su, WW. Naringin attenuates enhanced cough, airway hyperresponsiveness and airway inflammation in a guinea pig

model of chronic bronchitis induced by cigarette smoke. *International Immunopharmacology,* 2012 13, 301-307.

[79] Ma, Q; Xie, H; Li, S; Zhang, R; Zhang, M; Wei, X. Flavonoids from the pericarps of *Litchi chinensis. Journal of Agricultural and Food Chemistry,* 2014 62, 1073-1078.

[80] Manach, C; Morand, C; Gil-Izquierdo, A; Bouteloup-Demange, C; Rémésy, C. Bioavailability in humans of the flavanones hesperidin and narirutin after the ingestion of two doses of orange juice. *Europian Journal of Clinical Nutrition*, 2003 57, 235-242.

[81] Manach, C; Williamson, G; Morand, C; Scalbert, A; Remesy, C. Bioavailability and bioefficacy of polyphenols in humans. I. Review of 97 bioavailability studies. *The American Journal of Clinical Nutrition,* 2005 81, 230S-242S.

[82] McIntosh, CI; Mansell, RL. Three-dimensional distribution of limonin, limonoate A-ring monolactone, and naringin in the fruit tissues of three varieties of *Citrus paradisi. Journal of Agricultural and Food Chemistry,* 1997 45, 2876-2883.

[83] Mojzisová, G; Mirossay, L; Kucerová, D; Kyselovic, J; Mirossay, A; Mojzis, J. Protective effect of selected flavonoids on *in vitro* daunorubicin-induced cardiotoxicity. *Phytotherapy Research,* 2006 20, 110-114.

[84] Mojzisová, G; Sarisský, M; Mirossay, L; Martinka, P; Mojzis, J. Effect of flavonoids on daunorubicin-induced toxicity in H9c2 Cardiomyoblasts. *Phytotherapy Research,* 2009 23, 136-139.

[85] Mullen, W; Archeveque, MA; Edwards, CA; Matsumoto, H; Crozier, A. Bioavailability and metabolism of orange juice flavanones in humans: impact of a full-fat yogurt. *Journal of Agricultural and Food Chemistry,* 2008 56, 11157–11164.

[86] Mulvihill, EE; Allister, EM; Sutherland, BG; Telford, DE; Sawyez, CG; Edwards, JY; Markle, JM; Hegele, RA; Huff, MW. Naringenin prevents dyslipidemia, apolipoprotein B overproduction, and hyperinsulinemia in LDL receptor-null mice with diet-induced insulin resistance. *Diabetes,* 2009 58, 2198-2210.

[87] Mulvihill, EE; Assini, JM; Sutherland, BG; DiMattia, AS; Khami, M; Koppes, JB; Sawyez, CG; Whitman, SC; Huff, MW. Naringenin decreases progression of atherosclerosis by improving dyslipidemia in high-fat-fed low-density lipoprotein receptor-null mice. *Arteriosclerosis Thrombosis and Vascular Biology,* 2010 30, 742-748.

[88] Muthaiah, VP; Venkitasamy, L; Michael, FM; Chandrasekar, K; Venkatachalam, S. Neuroprotective role of naringenin on carbaryl induced neurotoxicity in mouse neuroblastoma cells. *Journal of Pharmacology and Pharmacotherapeutics,* 2013 4,192-197.

[89] Nagel, E; Meyer zu Vilsendorf, A; Bartels, M; Pichlmayr, R. Antioxidative vitamins in prevention of ischemia/reperfusion injury. *International Journal for Vitamin and Nutrition Research*, 1997 67, 298-306.

[90] Oršolić, N; Gajski, G; Garaj-Vrhovac, V; Dikić, D; Prskalo, ZŠ; Sirovina, D. DNA-protective effects of quercetin or naringenin in alloxan-induced diabetic mice. *European Journal of Pharmacology,* 2011 656, 110-118.

[91] Pandey, RP; Sohng, JK. Genetics of flavonoids. In: Ramawat KG; Merillon, JM, editors. Natural Products phytochemistry, botany and metabolism of alkaloids, phenolics and terpenes. *Heidelberg: Springer;* 2013; pp 1619-1641.

[92] Pavlović, AV; Dabić, DČ; Momirović, NM; Dojčinović, BP; Milojković-Opsenica, DM; Tešić, ZLj; Natić, MM. Chemical composition of two different extracts of berries harvested in Serbia. *Journal of Agricultural and Food Chemistry,* 2013 61, 4188-4194.

[93] Pickup, JC; Chusney, GD; Thomas, SM; Burt, D. Plasma interleukin-6, tumour necrosis factor-α and blood cytokine production in type 2 diabetes. *Life Science,* 2000 67,291-300.

[94] Priscilla, DH; Roy, D; Suresh, A; Kumar, V; Thirumurugan, K. Naringenin inhibits α-glucosidase activity: a promising strategy for the regulation of postprandial hyperglycemia in high fat diet fed streptozotocin induced diabetic rats. *Chemico-Biological Interactions,* 2014 210, 77-85.

[95] Pu, P; Gao, DM; Mohamed, S; Chen, J; Zhang, J; Zhou, XY; Zhou, NJ; Xie, J; Jiang, H. Naringin ameliorates metabolic syndrome by activating AMP-activated protein kinase in mice fed a high-fat diet. *Archives of Biochemistry and Biophysics,* 2012 518, 61-70.

[96] Qin, CX; Chen, X; Hughes, RA; Williams, SJ; Woodman, OL. Understanding the cardioprotective effects of flavonols: discovery of relaxant flavonols without antioxidant activity. *Journal of Medicinal Chemistry*, 2008 51,1874-1884.

[97] Qin, L; Jin, L; Lu, L; Lu, X; Zhang, C; Zhang, F; Liang, W. Naringenin reduces lung metastasis in a breast cancer resection model. *Protein Cell.* 2011 2, 507-516.

[98] Rajadurai, M; Prince, PS. Preventive effect of naringin on cardiac mitochondrial enzymes during isoproterenol-induced myocardial infarction in rats: a transmission electron microscopic study. *Journal of Biochemistry and Molecular Toxicology,* 2007 21, 354-361.

[99] Rajadurai, M; Stanely Mainzen Prince P. Preventive effect of naringin on lipid peroxides and antioxidants in isoproterenol-induced cardiotoxicity in Wistar rats: biochemical and histopathological evidences. *Toxicology,* 2006 228, 259-268.

[100] Raza, SS; Khan, MM; Ahmad, A; Ashafaq, M; Islam, F; Wagner, AP; Safhi, MM; Islam, F. Neuroprotective effect of naringenin is mediated through suppression of NF-κB signaling pathway in experimental stroke. *Neuroscience,* 2013 230, 157-171.

[101] Reddy, TK; Nagaraju, I; Kumar, KH; Lokanatha, V; Reddy, CD; Jagetia, GC. Cardioprotective effect of naringin in mice treated with doxorubicin. *Planta Medica,* 2008 74, 49.

[102] Richard, AJ; Amini-Vaughan, Z; Ribnicky, DM; Stephens, JM. Naringenin inhibits adipogenesis and reduces insulin sensitivity and adiponectin expression in adipocytes. *Evidence-Based Complementary and Alternative Medicine,* 2013 2013, 549750.

[103] Sabarinathan, D; Mahalakshmi, P; Vanisree, AJ. Naringenin, a flavanone inhibits the proliferation of cerebrally implanted C6 glioma cells in rats. *Chemico-Biological Interactions.* 2011 189, 26-36.

[104] Santos, CNS; Koffas, M; Stephanopoulos, G. Optimization of a heterologous pathway for the production of flavonoids from glucose. *Metabolic Engineering* 2011 13, 392-400.

[105] Schwinn, KE; Davies, KM. Flavonoids. In: Davies KM, editor. Plant pigments and their manipulation. Oxford: *Blackwell Publishing Ltd;* 2004; pp92-149.

[106] Sesso, HD; Buring, JE; Christen, WG; Kurth, T; Belanger, C; MacFadyen, J; Bubes, V; Manson, JE; Glynn, RJ; Gaziano, JM. Vitamins E and C in the prevention of

cardiovascular disease in men: the Physicians' Health Study II randomized controlled trial. *The Journal of the American Medical Association,* 2008 300, 2123-2133.

[107] Shi, Y; Dai, J; Liu, H; Li, RR; Sun, PL; Du, Q; Pang, LL; Chen, Z; Yin, KS. Naringenin inhibits allergen-induced airway inflammation and airway responsiveness and inhibits NF-kappaB activity in a murine model of asthma. *Canadian Journal of Physiology and Pharmacology,* 2009 87, 729-735.

[108] Shin, YW; Bok, SH; Jeong, TS; Bae, KH; Jeoung, NH; Choi, MS; Lee, SH; Park, YB. Hypocholesterolemic effect of naringin associated with hepatic cholesterol regulating enzyme changes in rats. *International Journal for Vitamin and Nutrition Research,* 1999 69, 341-347.

[109] Shulman, M; Cohen, M; Soto-Gutierrez, A; Yagi, H; Wang, H; Goldwasser, J; Lee-Parsons, CW; Benny-Ratsaby, O; Yarmush, ML; Nahmias, Y. Enhancement of naringenin bioavailability by complexation with hydroxypropyl-β-cyclodextrin. *PLoS One,* 2011 6,e18033.

[110] Stevens, JF; Taylor, AW; Deinzer, ML. Quantitative analysis of xanthohumol and related prenylflavonoids in hops and beer by liquid chromatography-tandem mass spectrometry. *Journal of Chromatography A,* 1999 832, 97-107.

[111] Subramanian, P; Arul, D. Attenuation of NDEA-induced hepatocarcinogenesis by naringenin in rats. *Cell Biochemistry and Function,* 2013 31, 511-517.

[112] Sulfikkarali, N; Krishnakumar, N; Manoharan, S; Nirmal, RM. Chemopreventive efficacy of naringenin-loaded nanoparticles in 7,12-dimethylbenz(a)anthracene induced experimental oral carcinogenesis. *Pathology Oncology Research,* 2013 19, 287-296.

[113] Surya Sandeep, M; Sridhar, V; Puneeth, Y; Ravindra Babu, P; Naveen Babu, K. Enhanced oral bioavailability of felodipine by naringenin in Wistar rats and inhibition of P-glycoprotein in everted rat gut sacs *in vitro*. *Drug Development and Industrial Pharmacy,* 2014 40, 1371-1377.

[114] Terao, J; Mukai, R. Prenylation modulates the bioavailability and bioaccumulation of dietary flavonoids. *Archives of Biochemistry and Biophysics*, 2014 559C, 12-16.

[115] Terra, X; Montagut, G; Bustos, M; Llopiz, N; Ardèvol, A; Bladé, C; Fernández-Larrea, J; Pujadas, G; Salvadó, J; Arola, L; Blay, M. Grape-seed procyanidins prevent low-grade inflammation by modulating cytokine expression in rats fed a high-fat diet. *The Journal of Nutritional Biochemistry,* 2009 20, 210-218.

[116] Testai, L; Martelli, A; Cristofaro, M; Breschi, MC; Calderone, V. Cardioprotective effects of different flavonoids against myocardial ischaemia/reperfusion injury in Langendorff-perfused rat hearts. The Journal of Pharmacy and Pharmacology, 2013 65, 750-756.

[117] Tomás-Barberán, FA; Clifford, MN. Flavanones, chalcones and dihydrochalcones-nature, occurrence and dietary burden. *Journal of the Science of Food and Agriculture,* 2000 80, 1073-1080.

[118] Torricelli, P; Ricci, P; Provenzano, B; Lentini, A; Tabolacci, C. Synergic effect of α-tocopherol and naringenin in transglutaminase-induced differentiation of human prostate cancer cells. *Amino Acids,* 2011 41, 1207-1214.

[119] Vafeiadou, K; Vauzour, D; Lee, HY; Rodriguez-Mateos, A; Williams, RJ; Spencer, JP. The citrus flavanone naringenin inhibits inflammatory signalling in glial cells and protects against neuroinflammatory injury. *Archives of Biochemistry and Biophysics,* 2009 484, 100-109.

[120] Vallejo, F; Larrosa, M; Escudero, E; Zafrilla, MP; Cerdá, B; Boza, J; García-Conesa, MT; Espín, JC; Tomás-Barberán, FA. Concentration and solubility of flavanones in orange beverages affect their bioavailability in humans. *Journal of Agricultural and Food Chemistry,* 2010 58, 6516-6524.

[121] Velozo, LSM; Da Silva, BP; Da Silva, EMB; Parente, JP. Constituents from the roots of *Bowdichia virgilioides*. Fitoterapia, 1999 70, 532-535.

[122] Wang, Y; Chen, S; Yu, O. Metabolic engineering of flavonoids in plants and microorganisms. *Applied Microbiology and Biotechnology,* 2011 91, 949-956.

[123] Winkel-Shirley, BSJ. Flavonoid biosynthesis. A colorful model for genetics, biochemistry, cell biology, and biotechnology. *Plant Physiology,* 2001 126, 485-493.

[124] Winkel-Shirley, BSJ. The biosynthesis of flavonoids. In: Grotewold E, editor. The science of flavonoids. *New York: Springer;* 2006; pp 71-95.

[125] Wu, H; Calvarano, M; Giacomo, A. Some flavanones in the peel of ten *Citrus* species and varieties in China. *Essenze-Derivati-Agrumari,* 1991 61, 103-112.

[126] Yang, W; Ma, J; Liu, Z; Lu, Y; Hu, B; Yu, H. Effect of naringenin on brain insulin signaling and cognitive functions in ICV-STZ induced dementia model of rats. *Neurological Sciences,* 2014 35, 741-751.

[127] Yannai, S. Dictionary of food compounds with CD-ROM: Additives, flavors, and ingredients. Boca Raton, FL: Chapman & Hall/CRC; 2004.

[128] Yi, LT; Liu, BB; Li, J; Luo, L; Liu, Q; Geng, D; Tang, Y; Xia, Y; Wu, D. BDNF signaling is necessary for the antidepressant-like effect of naringenin. *Progress in Neuro-psychopharmacology and Biological Psychiatry,* 2014 48, 135-141.

[129] Yilma, AN; Singh, SR; Morici, L; Dennis, VA. Flavonoid naringenin: a potential immunomodulator for Chlamydia trachomatis inflammation. *Mediators of Inflammation*, 2013 2013:102457.

[130] Yoon, H; Kim, TW; Shin, SY; Park, MJ; Yong, Y; Kim, DW; Islam, T; Lee, YH; Jung, KY; Lim, Y. Design, synthesis and inhibitory activities of naringenin derivatives on human colon cancer cells. *Bioorganic and Medicinal Chemistry Letters.* 2013 23, 232-238.

[131] Yousuf, S; Sudha, N; Murugesan, G; Enoch, IV. Isolation of Prunin from the fruit shell of *Bixa orellana* and the effect of β-cyclodextrin on its binding with calf thymus DNA. *Carbohydrate Research*, 2013 365, 46-51.

[132] Zhang, FY; Du, GJ; Zhang, L; Zhang, CL; Lu, WL; Liang, W. Naringenin enhances the anti-tumor effect of doxorubicin through selectively inhibiting the activity of multidrug resistance-associated proteins but not P-glycoprotein. *Pharmaceutical Research,* 2009 26, 914-925.

INDEX

B

C

D

E

F

G

H

I

J

K

L

M

N

O

P

Q

R

S

T

U

V

W

X